From Data to Insights

This book offers a clear and accessible guide to cross-tabulation analysis, transforming a complex subject into an accessible topic. It diverges from traditional statistical texts, adopting a conversational tone that addresses common questions and concerns. The author demystifies intricate concepts, with clear explanations and relatable analogies that make the material approachable for readers with varying levels of mathematical expertise.

Unique in its approach, the book avoids overwhelming readers with complex formulas and instead focuses on the principles underlying cross-tabulation analysis. This method ensures that the content is applicable regardless of specific statistical software used, making it a versatile resource.

Targeted at a diverse audience, the book covers the spectrum from foundational elements to comparatively more advanced topics in cross-tabulation analysis. It includes a comprehensive glossary and an appendix of detailed examples, providing practical insight and aiding understanding of key concepts. This book is an invaluable resource for students, researchers, and educators alike, offering a fresh perspective on cross-tabulation analysis that emphasises clarity and practical application.

Key Features:

- Employs a conversational style, making complex statistical concepts in cross-tabulation analysis accessible and engaging for all readers.

- Combines minimal use of formulas with practical examples, ensuring easy comprehension and application, even for those with minimal mathematical background.

- Features a consistent running example for continuity, complemented by diverse real-world scenarios to solidify understanding of key concepts.

- Independently valuable without reliance on specific statistical software, emphasising fundamental principles for adaptability across various platforms.

- Progressively guides readers from foundational basics to comparatively more advanced methods, supplemented by a comprehensive glossary and detailed appendix for an enriched learning.

Gianmarco Alberti, Senior Lecturer at the University of Malta, specialises in applying data analysis across diverse fields, from archaeology to criminology. His pedagogical expertise spans various levels, including Bachelor's, Master's, and Doctoral programmes, in multiple disciplines and institutions. His contributions include developing open-access software for spatial statistics and multivariate analysis, and authoring a substantial body of scholarly work, including a monograph and scholarly articles in renowned journals.

From Data to Insights

A Beginner's Guide to
Cross-Tabulation Analysis

Gianmarco Alberti

CRC Press
Taylor & Francis Group
Boca Raton London New York

CRC Press is an imprint of the
Taylor & Francis Group, an **informa** business

A CHAPMAN & HALL BOOK

Designed cover image: © Gianmarco Alberti

First edition published 2025
by CRC Press
2385 NW Executive Center Drive, Suite 320, Boca Raton FL 33431

and by CRC Press
4 Park Square, Milton Park, Abingdon, Oxon, OX14 4RN

CRC Press is an imprint of Taylor & Francis Group, LLC

© 2025 Gianmarco Alberti

ISBN: 9781032720388 (hbk)
ISBN: 9781032726304 (pbk)
ISBN: 9781032726328 (ebk)

DOI: 10.1201/9781032726328

Typeset in Palatino
by Deanta Global Publishing Services, Chennai, India

Contents

Foreword

The analysis of categorical data is pervasive, particularly in the current era of big data. Surveys of all kinds collect data that occur naturally as a category or exist as numerical data waiting to be intervalised. It should therefore come as no surprise that the tools needed to analyse categorical data, and in particular cross-tabulations (a pivotal way of summarising such data), are extremely important and are needed across diverse domains, such as archaeology, education, social science, economics, behavioural science, criminology, biomedical and data science.

There are plenty of books out there that describe a plethora of ways in which cross-tabulations can be analysed, although most of them delve into the technical and heavily statistical features. This often leaves some readers bewildered and therefore at a loss as to how and why their data need to be analysed. This book gives an excellent overview of the basics of analysing cross-tabulations and does so in a friendly and engaging way. The author not only discusses a variety of tools necessary for initiating the analysis of cross-tabulations, but he does so with great care and attention. While steering clear of the technical rigour that often dominates discussions in most books on the topic, this book provides a more accessible approach to studying cross-tabulations. The statistical vernacular is kept to a minimum, while the applications and explanations are discussed in a way that should engage all readers.

This book introduces the reader to the analysis of cross-tabulations and, more generally, categorical data, in a very natural and structured way.

Chapter 1 starts with a discussion of what categorical data and cross-tabulation are and why analysing one is important. This is done with plenty of simple examples with real-world context. Attention in this chapter, and throughout the book, is focused on how to analyse one and two sets of categories, thereby keeping the descriptions grounded in understanding the basics. Therefore, the reader will find that this book avoids much of the mathematical complexities that often accompany descriptions of how to analyse categorical data.

Chapter 2 describes how to gain an understanding of the relationship, or association, between the two sets of categories that form a cross-tabulation. It starts by providing an intuitive description of independence. This is important because determining whether there is an association between different sets of categories is the foundation on which much of categorical data analysis rests. The reader is then directed to how best to assess whether the data exhibits independence by providing a practical and non-technical description of the chi-squared test. This chapter ends with an explanation of how to understand statistical significance, including a discussion of the p-value, a crucial statistical concept that plays a big role in all areas of statistics, including categorical data analysis and the study of cross-classifications.

All statistical techniques have their pros and cons and the tools used to analyse cross-tabulations are no exception to this rule. So, Chapter 3 provides further insight into the pros and cons of using the chi-squared test that is described in Chapter 2. This chapter should prove enlightening as the reader will gain a deeper appreciation for when this test can be used and the necessary cautions attached to it (cautions that are very often overlooked when the chi-squared test is used in a real-world context).

As Chapters 2 and 3 provide insight into the chi-squared test and its pros and cons, Chapter 4 discusses other ways of determining the association between two sets of categories that form a cross-tabulation. There are hundreds of methods that can be used to do this, but the author rightly only provides a description of the key measures of association such as the odds ratio, Cramér's measures, and the phi-statistic (among others). These descriptions are all done with a practical view of categorical data analysis. An excellent inclusion in this chapter is a description and application of the Goodman–Kruskal lambda statistic, a measure of association that, unlike the chi-squared statistic, gives the researcher the tool to analyse the causal link between their sets of categories (much like simple linear regression provides a causal link between numerical data).

While the author takes us on a trip through the many ways that a cross-tabulation formed from two sets of categories can be analysed, Chapter 5 provides some guidance on how three sets of categories can be analysed. This is done by highlighting the various ways in which association (described in Chapter 2) can be assessed and includes a practical discussion of the famous Simpson's paradox for analysing stratified data. Chapter 5 also provides a practical guide to using the Cochran–Mantel–Haenszel and Breslow–Day tests for a cross-tabulation formed using three sets of categories.

Chapter 6 gives a global view of all of the tools discussed in the earlier chapters by showing how they can all be applied to a single dataset comprising three sets of categories.

Finally, the book concludes by offering readers a glimpse into the expansive world of analysing cross-tabulations (Chapter 7). A brief description of log-linear models and logistic regression is provided, which gives the reader more flexibility in how they wish to analyse their categorical data. The author also provides a glimpse into correspondence analysis, a tool designed for visualising the structure of the association between two or more sets of categorical variables. The same chapter provides a detailed bibliography that offers an excellent list of essential readings for each of the topics described in the book.

Gianmarco Alberti is world renowned as an excellent practitioner and communicator of categorical data analysis, especially with a practical flavour. Therefore, we feel that the reader of this book will gain a lot from how well he has described and structured the topics covered in this book. All of the tools and tips described within these pages are highly practical, requiring only a fundamental knowledge of statistics. They should therefore prove valuable to all who immerse themselves in this excellent book.

Eric J. Beh
National Institute for Applied Statistics Research Australia (NIASRA),
University of Wollongong, Australia, & Centre for Multi-Dimensional Data
Visualisation (MuViSU), Stellenbosch University, South Africa

Rosaria Lombardo
Department of Economics, University of Campania "Luigi Vanvitelli",
Capua (CE), Italy

In the vast landscape of research, where numbers and statistics often act as a barrier to understanding, this book comes forth as a helpful guide, simplifying the complex subject of cross-tabulation analysis. As a Master's student exploring various research methods, and both quantitative and qualitative strategies, I have found solace and enlightenment within the pages of this remarkable book.

One of its primary objectives is to make this seemingly intimidating subject accessible and understandable to a wider readership. It is with this goal in mind that the book unfolds, offering a clear roadmap for those beginners seeking clarity in cross-tabulation analysis.

As someone who once grappled with the intricacies of numbers and statistical notions, I can attest to the transformative power of Dr Alberti's teachings. His approach goes beyond mere instructions; it is a journey of skill development. This book, crafted with precision and empathy, serves as a friendly companion, ushering us from confusion to comprehension.

What sets Dr Alberti apart is his keen awareness of the challenges faced by students. This book, a testament to his commitment, is a gift to those who have felt the daunting weight of statistical uncertainty. It not only imparts knowledge but also instils a sense of confidence in the readers. The chapters unfold like a well-crafted narrative, gradually unravelling the particulars of cross-tabulation analysis in a manner that is both engaging and enlightening.

As a Master's student who has traversed the chapters of this book, I extend my heartfelt gratitude to the author. His unwavering dedication has left an indelible mark on my academic journey. I am confident that this book will be an invaluable guide for anyone delving into cross-tabulation analysis, providing the necessary tools to not only understand but also appreciate the reasoning behind it.

Erica Micallef Filletti
Criminologist & Master of Arts in Criminology student
University of Malta

Preface

My journey into the world of research began out of a profound need to comprehend and use categorical data effectively. More than just an intellectual pursuit, this understanding was a crucial foundation for my research aims. As I navigated this domain, the importance of a robust grounding in data analysis fundamentals stood out clearly.

As a student, certain books fascinated me; not necessarily the thickest or the most referenced ones. Instead, they had a unique trait: the authors had a remarkable gift for making intricate ideas feel simple without effort. These were not just books; they were catalysts for curiosity and joy in the learning process. I have always hoped to channel this same clarity and simplicity into my own teachings and writings.

I drafted this book during the nights of the second semester of the academic year 2022–2023. It was during these quiet moments that the dialogue within these pages began to take shape. Picture it: me, imagining a student from one of my courses sitting across from me, trying to address their potential doubts and curiosities about data. This is why you will find sections of the book that resemble a dialogue; it is a simulated conversation, addressing real concerns that often bubble up in my actual classroom discussions. These imagined, yet remarkably real, dialogues were not born in isolation; they were cultivated from genuine interactions and recurrent themes observed in my daytime (and sometimes evening) classroom.

As a senior lecturer specialising in research methods and quantitative strategies across various fields, I often noticed a trend. Many students, particularly from the social sciences, showed a tangible fear of numbers and statistical notions. This was not just an occasional observation; it highlighted a deeper issue regarding the way quantitative data is generally not only received from but also passed on to students. I have always believed in the power of engagement in learning. Intricate ideas, when distilled and relayed in familiar terms, can turn from intimidating terminologies into captivating concepts. So, how can one make these abstract, often daunting, topics more relatable? With a study background in Classics that took me from Hesiod to Constantine the Great, with all that lies in between, I have come to respect the nuances and power of language. That is one reason you will find analogies sprinkled throughout this book, with the aim of making the abstract more tangible. It is not so much about simplifying the subject but uplifting the learner.

Reading through these pages, statisticians and data analysts may perceive the prose to be sometimes verbose or overly detailed. This is a deliberate choice. Drawing upon my own experiences as a student, and from observing others in their learning journey, I recognise that complexity may often generate confusion. Therefore, I have endeavoured to anticipate your queries, elucidate potential areas of misunderstanding, and, at times, intentionally circle back to reiterate key concepts, especially those that tend to be more challenging to grasp. The repetition is not merely to fill pages, but to underscore significance, solidify understanding, and provide an implicit check-in with your learning as we progress through the material.

In this book, you will find a distinctive absence of reference to specific statistical software, an intentional choice guided by multiple considerations. Firstly, the rapid evolution and updates of software technology inherently risk rendering any detailed, software-specific, instructions quickly outdated and potentially obsolete. However, the second, and perhaps more crucial reason, centres on an educational philosophy: I do believe that the mastery of statistics and data analysis transcends mere familiarity with a particular software's operations. This conviction was further solidified by a personal experience. Once, while discussing the chi-squared test with a colleague, they casually remarked, 'Ah, yes, I know how to do it in [name of a software here]'. This interaction exemplified a mindset I have observed pretty often among students as well as instructors: the conflation of statistical knowledge with software proficiency. It is a perspective I intend to challenge and reshape.

Throughout the years, I have encountered other numerous instances where students and colleagues alike equate data analysis with the navigation of specific, and sometimes intricately complex, software. The intention behind this exclusion is not to understate the value of software in facilitating data analysis. Rather, it seeks to stress a key principle: that a clear and robust understanding of the underlying rationale and methodologies of cross-tabulation analysis should take precedence. With this foundation, analysts are empowered to utilise any software with which they feel most comfortable, thereby applying the methodologies and approaches elucidated in this book in a more versatile manner.

With these goals in mind, this book aims to be more than a mere guide. It is crafted to hopefully bridge the gap between uncertainty and comprehension, between ambiguity and lucidity. In our increasingly data-driven society, grasping and deciphering data is not just a commendable skill; it is essential. I hope this book reshapes the perception of cross-tabulation analysis from being an overwhelming hurdle to an engaging and fruitful journey.

Now that the book is complete, I wish to extend my gratitude to my students across different departments and faculties at the University of Malta. Your candidness in expressing doubts and apprehensions concerning numerical data, and sharing your perspectives on intricate concepts, has been invaluable. Your insights have not only enriched this book but also deepened my understanding of the pedagogical challenges and opportunities in this field.

I want to express my heartfelt thanks to many colleagues and friends who took the time to read earlier drafts of this book. Whether through a glance or a deep dive, your feedback, comments, suggestions, and words of encouragement have been instrumental in refining this work. In alphabetical order, I am indebted to Eric J. Beh, Rachel Bennett, John Betts, Luke J. Buhagiar, Miriam Farrugia, Saviour Formosa, Janice Formosa Pace, Reuben Grima, Rosaria Lombardo, Gillian Martin, Erica Micallef Filletti, Harry Olieman, Gordon Sammut, Sandra Scicluna, Davide Tanasi, and Nicholas C. Vella. The varied perspectives of each one of you have greatly contributed to the development of this work.

I would like to specially acknowledge Reuben Grima for reading one of the very first drafts of this work and providing feedback about its readability, clarity, and flow. His early and valuable insights were fundamental in guiding the subsequent versions of the book.

I would also like to extend a special note of gratitude to Eric J. Beh and Rosaria Lombardo. Amidst their scholarly busy schedules, they found the time to read this work and draft a foreword that is both insightful and refreshing. Both are esteemed experts in the field of categorical data analysis and advanced methods applied to such data, with an extensive list of publications in these areas. Their kind contribution enriches this book and brings a depth of professional expertise.

A heartfelt acknowledgement also goes to Erica Micallef Filletti, currently reading for her Master of Arts in Criminology. Her foreword definitely provides novice readers with a relatable entry point into the subject. Also, her thoughtful feedback on some draft chapters was particularly insightful. Erica's ability to pinpoint areas in need of clarification not only greatly enhanced the overall clarity of this book but also added a unique and invaluable perspective.

I want to thank the professionals involved in the publication process. Their expertise and support were crucial in transforming an idea into a tangible reality. Lara Spieker, Editor for Statistics and Data Science at Chapman & Hall/CRC Press, guided the project with exceptional insight and dedication. Curtis Hill, Editorial Assistant for Statistics at the same publishing house, provided invaluable assistance and attention to detail. Rachel Cook, as Senior Project Manager at Deanta, skilfully managed the many moving parts of the book's production with efficiency and care. Their kindness and helpfulness at every stage were instrumental in making this book what it is.

I express my heartfelt thanks to the anonymous copyeditor whose keen eye for detail and dedication to clarity greatly enhanced the final manuscript. Their meticulous work in editing the text, picking up typos, and correcting errors that my tired eyes overlooked was invaluable.

Last but not least, I must thank the reviewers of the book, whose feedback on the first draft was indispensable in refining and strengthening the manuscript. Any errors or shortcomings within this work are solely my responsibility.

To close this preface, my desire is for this book to resonate with and aid a broad spectrum of readers. Whether you are a student, a researcher, or merely intrigued by cross-tabulations, this book is for you. I have tried to dispel the usual anxieties tied to complex mathematical representations, aiming instead for transparency, understanding, and possibly even an interest in the art of data analysis. I invite you to turn these pages not with hesitation but with the zest of exploration, and the assurance that knowledge is accessible to all.

Happy reading!

1

Cross-Tabulations

1.1 Introduction

If you are reading this book, it is likely that you are dealing (or plan to deal) with data featuring categorical variables, and that you are looking for insights into how to approach their analysis. Categorical data are everywhere in the social sciences, as well as in other fields of research, and one cannot escape making sense of tables, rows, columns, percentages, ratios, odds ratios, chi-squared test, measures of categorical association, and the like.

Before we proceed, there is something essential to highlight: this book is a deviation from the norm. While most statistics books dive into a sea of complex distributions, probabilities, and intimidating formulae, expecting you to swim with a robust statistical and mathematical background, this one does not. Instead, it firmly anchors itself on the practical aspects of cross-tabulation. Rather than pushing you into deciphering intimidating equations from the outset, we will start our journey by understanding the very essence of cross-tabulations. The goal? To empower you with the ability of making sense of data without getting lost in the mathematical labyrinth.

Throughout this book, I will introduce concepts gradually and I will use a colloquial language to make sure that you do not need a PhD in statistics to understand what I am saying. I will avoid technical jargon as much as possible to keep things simple and accessible. The aim is to provide clear frameworks where simplified, yet relevant, information can find its place, so that readers can make their first step into the world of categorical data analysis with ease. I want to make sure that you are comfortable with the material, so you can learn at your own pace and have fun along the way. Humour will be sprinkled throughout to ensure an enjoyable read.

It is crucial to note that, while this book is designed to be easily digestible, it does not mean the journey will be effortless. True understanding often requires revisiting concepts, rereading sections, and scrutinising tables and explanations multiple times. Just as mastering any skill demands practice and patience, grasping the nuances of categorical data analysis may necessitate revisiting certain chapters or examples. It is not a sign of inadequacy,

DOI: 10.1201/9781032726328-1

but an indication of genuine engagement. If you find yourself going back to a particular section or concept, know that it is a part of the learning process.

My goal is to provide you with a clear understanding of how we can extract valuable information and uncover hidden patterns from data presented in table format. In the upcoming pages, I will be using the well-known Titanic dataset as a running example to elucidate the underlying principles of cross-tabulation analysis. The Titanic data is a classic dataset often used in statistics and data analysis training. It comprises information about the passengers on the ill-fated voyage of the RMS Titanic, including details such as age, gender, passenger class, and survival, among others. The reason for employing this particular dataset is manifold. Firstly, the dataset is publicly available on the web (see Section 7.4.1), making it a convenient choice for readers to access and engage with the examples discussed in this work. Secondly, it has been incorporated in various statistics packages as an in-built dataset, highlighting its pedagogical utility.

Moreover, the dataset contains a mix of categorical variables, thereby providing a rich context for demonstrating various concepts around categorical data analysis. It allows for a comprehensive exploration of cross-tabulations, chi-squared tests, measures of association, and more, in a real-world setting. By leveraging a dataset that is both well known and representative of common challenges faced in categorical data analysis, it aims to bridge the gap between theoretical concepts and practical understanding. The Titanic dataset, with its historical significance and straightforward variable structure, serves as an ideal candidate for breaking down complex ideas into digestible, relatable examples. Some fictitious (yet realistic) datasets are used in Chapter 6, where the key concepts reviewed throughout this work are put into practice with new data.

To generate most of the tables featured in this book I used the *Jamovi* statistical software, which is free and user-friendly. For other images (namely, the ones in Chapter 3 showing different types of chi-squared residuals and mosaic plots), I have used the *chisquare* and the *vcd* R packages, respectively. I provide you with more information on them in Chapter 7.

In Chapter 1, you will learn about the foundational understanding of cross-tabulations, crucial for analysing categorical data, distinguishing between various types of variables, and constructing a 2 × 2 table using a simplified fictitious Titanic dataset as an example.

In Chapter 2, you will delve deeper into cross-tabulation analysis, introducing the concept of independence and statistical significance and exploring the relationship between pairs of variables using the chi-squared test and the associated *p*-value.

Chapter 3 expands on the chi-squared test, introducing tools like standardised residuals and mosaic plots. The chapter also tackles the challenges of small expected frequencies, in both small and larger tables, discussing approaches such as pooling levels, applying the $(N-1)/N$ correction, and utilising Fisher's test. It also explores computational techniques such as permutation and Monte Carlo methods as alternative solutions when the chi-squared test's applicability is in question.

In Chapter 4, you will explore various measures of association to quantify the strength of dependence between pairs of variables, learning about both chi-square-based and non-chi-square-based measures. The chapter examines the pros and cons of those measures and also touches upon the issue of how to verbally articulate the magnitude of the association that they express.

Chapter 5 introduces you to stratified cross-tabulations, exploring deeper insights into variable relationships and introducing concepts like Simpson's paradox, conditional independence, and homogeneous association through tests like the Cochran–Mantel–Haenszel and the Breslow–Day.

In Chapter 6, we will undertake a new complete analysis using a fictitious survey example, employing the various approaches reviewed in the preceding chapters to explore associations and providing guidance on formally reporting analysis results.

Chapter 7 introduces the broad landscape of cross-tabulation analysis beyond 2 × 2 tables, touching upon advanced techniques like log-linear modelling, logistic regression, and correspondence analysis. The chapter also provides a curated list of readings to deepen your understanding of cross-tabulation analysis.

As you progress through each chapter, you will find concise sets of key takeaways at the end, designed to summarise the most crucial points and help solidify your understanding of the topics covered. These takeaways serve as quick references for revisiting the material or for last-minute prep before putting your knowledge into practice.

To further enhance your learning experience, an Appendix provides practical, hands-on examples, particularly focusing on the application of odds and odds ratios, to reinforce the concepts and calculations encountered in Chapter 4. This additional section is designed to solidify your understanding through practical application, ensuring that the theoretical knowledge gained throughout the book is not only retained but also skilfully applied.

Finally, to support your journey through cross-tabulation analysis, a comprehensive glossary of terms is provided. This glossary acts as a quick reference guide, assisting you in grasping the terminology and concepts essential for mastering the subject matter.

Overall, while the chapters offer in-depth insights, the key takeaways, the Appendix, and the glossary further enrich your learning experience, making the subject approachable and (hopefully) easier to digest.

1.2 Defining the Book's Boundaries

While this book provides an introduction to the analysis of cross-tabulations for categorical variables, it is important to address expectations regarding the depth of coverage in certain areas. The focus is squarely on what we will soon define as *nominal* categorical variables (see Section 1.4). The decision to not delve into *ordinal* categorical variables is deliberate for several reasons. One key issue is that ordinal variables make the interpretation of a cross-tabulation more nuanced. Unlike nominal variables, ordinal ones have a natural order, adding an extra layer of complexity when one tries to understand their relationship. More importantly, commonly used statistical tests and measures discussed in this book are not designed to capture the ordered nature of ordinal variables. These tests and measures are effective when applied to nominal variables, as they do not consider any inherent ordering. Therefore, to maintain focus, I have chosen to limit the discussion to nominal categorical variables.

I will focus on 2 × 2 tables, which feature two rows and two columns. While almost all the methods I discuss can be applied to larger tables (more on this in Chapter 7), I chose to focus on 2 × 2 tables as they provide an easier way to illustrate concepts. They form the foundation of understanding larger, more intricate tables. As the saying goes, before riding a bicycle, one must learn to walk. However, it is crucial to clarify that my emphasis on small tables is purely for illustrative purposes and not a recommendation to amalgamate categories. Under no circumstances I advocate for the simplification of data by merging categories in order to force a table into a 2 × 2 format. This can in fact lead (among other things) to an unduly loss of valuable information (see Chapter 7).

Moreover, this book will not delve into the computation and interpretation of confidence intervals around odds ratios. This decision has been made to avoid overwhelming readers, particularly those new to the field, with additional complex concepts. However, it is crucial to acknowledge the significance of these intervals in statistics. In this work, we will focus on employing a number of tests to determine the existence of an association between variables (Chapters 2 and 5), and will use the odds ratio(s) (along with different coefficients of association) as a follow-up measure(s) to quantify the strength and direction of any identified dependence (Chapter 4).

1.3 How to Use This Book

This book has been designed to cater to a diverse audience, ranging from students and professionals new to cross-tabulation analysis to those with some experience in the field looking to deepen their understanding. To help you navigate the content and tailor your learning experience, here is a guide on how to use this book.

For Beginners:
If you are new to cross-tabulations, it is advisable to start from Chapter 1 and proceed sequentially through the book. Each chapter builds on the previous

one, introducing new concepts and methods in a progressive and digestible manner. Engage actively with the example analyses, and consider summarising key concepts in your own words or creating mind maps to visualise connections. The key takeaways at the end of each chapter and the comprehensive glossary are invaluable resources, not just for consolidation but also for quick reference as you progress. Additionally, the Appendix provides practical, hands-on examples, particularly focusing on the application of odds and odds ratios, reinforcing the concepts and calculations you will encounter in Chapter 4.

For Intermediate Learners:
If you already have a foundational understanding of categorical data analysis, you might opt to skim through the initial chapters, focusing more on areas where you wish to deepen your knowledge. Utilise the chapter summaries and key takeaways to swiftly identify sections most pertinent to your needs. Engage with the example analyses, challenging yourself to apply the discussed methods and, if possible, relate them to your own work or field of interest. Remember to use the glossary for quick clarifications and to ensure a solid grasp of terminology. The Appendix will be particularly useful in solidifying your understanding of odds and odds ratios, which are also crucial for advanced techniques such as logistic regression, discussed in Chapter 7.

For Advanced Users:
For those with a robust foundation in categorical data, this book can serve as a reference guide on cross-tabulation analysis. Navigate freely to specific chapters or sections that align with your interests or current needs. The example walkthroughs provide practical insights and can serve to complement and enhance your existing knowledge base. Additionally, Chapter 7 and the recommended readings at the end of the book are excellent resources, guiding you towards more advanced topics and further readings. The Appendix, with its detailed examples, can also aid in refining your approach to analyses and interpretations.

For Instructors:
This book can be a valuable resource in a teaching setting. The structured layout and logical progression of chapters provide a natural flow for a course on cross-tabulation analysis. Examples and walkthroughs can be integrated as in-class exercises, while the key takeaways and glossary serve as effective revision tools for students. Encourage your students to actively engage with the material, work in groups, and utilise the glossary and external resources to enrich their learning experience. The Appendix's scenarios, drawn from criminology, offer real-world contexts that can be used to challenge students and foster a deeper understanding of the material.

A Note to All Readers:

Regardless of your level of expertise or the purpose of your reading, remember that mastery is a journey that requires practice, patience, and perseverance. Engage actively with the material, replicate the examples, and do not hesitate to revisit concepts as needed. Learning is a personal and often communal journey; take advantage of the resources provided, both within and outside this book, and remember that the journey is as important as the destination. The Appendix is a testament to this, providing a space to reiterate and master the nuances of odds and odds ratios, ensuring a comprehensive grasp of these concepts before moving on to more complex analytical techniques.

1.4 The Building Blocks: Understanding Statistical Variables

Before progressing into the depths of cross-tabulations and the nuances of categorical data, it is important to provide you with a clear (and admittedly concise) description of statistical variables and of their classifications. Essentially, a statistical variable is a specific characteristic or quantity that has been observed or measured in each member (be it a person, an object, or anything we might be interested in studying) of the sample we study. Broadly categorised, these variables are either *quantitative* or *qualitative*. The latter are often referred to as *categorical*.

Quantitative variables are those that represent measurable quantities, providing numerical data that can be subjected to mathematical operations. They come in two flavours. The first is discrete data, which encompasses distinct and separate values. For instance, the exact number of books on a shelf, or students in a classroom, would be represented using discrete values. On the other hand, continuous data represent measurements that span a range, such as height, weight, or the time taken to complete a task. These variables can take on any value within a specified range.

Conversely, qualitative (or categorical) variables deal with non-numeric data: they express a quality that we consider either present or absent in each individual subject (or object) we study. Think of a classroom with a number of students; for every subject (student) we could take note of one or more qualities. For instance, their "hair colour", "eye colour", "preferred music genre", "nationality", and so forth. Each of those qualities corresponds to a categorical variable, representing a characteristic that (as said) we consider either present or absent in each student.

For example, when it comes to the variable "hair colour", having blond (or brown, or black) hair is a quality that a student either has or does not have. The same holds true for the other variables. These variables sort our observations (students, in our example) into specific groups, offering a descriptive nature rather than a measurable one. With categorical variables, what we can do is to count how many observations have specific characteristics, that is, fall into specific groups. For example, when it comes to the "hair colour" variable, we can count how many students in that classroom have blond, brown, or dark hair. The same applies for the "eye colour" variable: how many students have blue, brown, or dark eyes. By the same token, as for the "preferred music genre", how many of them listen to pop, rock, blues, or jazz music.

Within the realm of categorical variables, we have *nominal* variables, which simply classify observations without any inherent order, like the variable "type of fruit" or "nationality". Then, there are *ordinal* variables, possessing a distinct and implicit order, such as the "ranking" (poor, fair, good) or "education level" (primary, secondary, tertiary) variable.

Categorical variables are the building blocks of cross-tabulations.

1.5 Cross-Tab 101: The Basics Unveiled

What is a cross-tabulation (hereafter cross-tab) or a contingency table? To answer this question, let's have a look at Table 1.1 and imagine that we have collected information about 6 individuals.

Every row in that short list is an *observation* (a subject we are studying). For each observation, we record two attributes: the gender assigned at birth (either *female* or *male*) and having *survived* or *died* in the (in)famous Titanic shipwreck. Note that, when we use the term "gender" in this book's tables or text, we are referring to "gender assigned at birth".

The two attributes we are recording can be generically called *variables*, as discussed in Section 1.4. From this point on, variable names are capitalised to distinguish them from regular text and to emphasise their specific role within datasets and analyses. Those two variables are called (nominal)

TABLE 1.1

Table Summarising the SURVIVAL Outcomes of Six Individuals, Categorised by GENDER Assigned at Birth

Obs. No.	GENDER	SURVIVAL
1	Male	Died
2	Male	Survived
3	Female	Survived
4	Female	Survived
5	Female	Died
6	Male	Survived

categorical because (as said in the preceding section) they do not express a quantity that we can actually measure (like, for example, body weight, body temperature, IQ score, stature, distance, and the like), but a quality that we can consider as either being present or not in each individual subject we study. The two variables we are taking into account are therefore GENDER and SURVIVAL. In our example, each variable features two groups (aka *levels* in more technical jargon): *male–female* and *died–survived*, respectively.

In Table 1.1, the first individual is a male who unfortunately died. The second one is again a male who survived. The third observation is a female who survived; and so on and so forth. At this point, you are surely thinking: 'wait, you were supposed to explain what a cross-tab is, and you are talking about a list of people instead?'. That is what you're thinking, aren't you? Believe me: we are just one step away from our first small cross-tab. Bear with me.

From our list of observations (subjects), we can easily build our cross-tab by simply counting how many males died and how many males survived, and how many females died and how many females died. As simple as that. Have a look at Table 1.2.

If I were to use a more technical language, I should have said that to build our cross-tab we (well….our stats software) have to cross-tabulate GENDER against SURVIVAL (or the other way around, it does not really change things). In other words, we have to tabulate our two categorical variables against one another. The number of groups featuring the 2 categorical variables determines the size of the table. In this case, since we have 2 groups for GENDER and 2 groups for SURVIVAL, our table can be called a 2×2 (two by two) cross-tab. Needless to say, a cross-tab can be of any size, and the number of rows and columns can vary. In other words, do not expect to see either a 2×2, or a 3×3, or a 4×4 cross-tab; we can well have a table whose size is (say) 2×3, or 4×3, or 5×10, or any combination of row and column numbers.

Reading our table along columns and starting from the left, we see that (in our small sample of 6 individuals) we have 1 male who died and 2 who survived; moving to the adjacent column, we see that we have 1 female who died and 2 who survived. And that proves a handy summary of our dataset,

TABLE 1.2

Cross-Tabulation of SURVIVAL Outcomes and
GENDER for a Sample of Six Individuals

	GENDER		
SURVIVAL	**Male**	**Female**	**Total**
Died	1	1	2
Survived	2	2	4
Total	3	3	6

Note: Data from Table 1.1.

doesn't it? Also, let me draw your attention to some other details of this small cross-tab. Again, reading the table along columns, at the bottom we see that we have 3 males and 3 females in total. Those are called column *sums* (or *marginals*, or *totals*). Reading the table along the rows (that is, horizontally), we see that we have a total of 2 individuals who died and 4 who survived (row *sums*, or *marginals*, or *totals*). The cell at the very bottom-right indicates the size of our sample; it tells us how many individuals overall our table cross-tabulates: 6 in total.

Now that you know the numbers and their meaning, let's give some more general names to those figures. The numbers indicating how many individuals fall at the intersections of the groups of the two variables are called *observed frequencies* (or *observed counts*). Why observed? Because they are what we *observe* in our sample, as simple as that. The column sums and the row sums are also called *marginal frequencies*. The number expressing the overall number of individuals in our sample (6 … remember?) is called the table's *grand total*. Let me give you just a little more information and I promise we will soon move on to another chapter.

Think about this: the column marginal frequencies give us an idea of how many males and females are in our sample, not taking into account SURVIVAL, that is, disregarding the distinction between who died and who survived. Regardless of who died or survived, we have (as seen earlier on) 3 males and 3 females. The row marginal frequencies give us an idea of how many individuals died or survived, ignoring GENDER. Regardless of who is male or female, we have (as seen) 2 dead and 4 survived persons in our sample. And that is all for now, as promised.

1.6 Key Takeaways

- Statistical variables are characteristics or quantities that are observed or measured.
- There are two main types of variables: quantitative (numeric) and qualitative (categorical).
- Quantitative variables can be discrete (distinct values) or continuous (range values).
- Categorical variables are further divided into nominal (without order) and ordinal (with a distinct order).
- Cross-tabulation (cross-tab) is a method to display the frequency of categorical variables against one another.
- A cross-tab can vary in size based on the number of groups in the analysed variables.

- Constructing a cross-tab involves counting how many observations fall at the intersection of each level of the two categorical variables under analysis (for example, female survivor, male survivor).
- Cross-tabs provide a summarised view of data, making patterns more evident.
- Elements in a cross-tab include observed frequencies (counts at intersections), marginal frequencies (column and row sums), and the grand total (total observations).

2

Cross-Tab Analysis and Introduction to the Chi-Squared Test

2.1 How to Look at Cross-Tabs

2.1.1 Exploring Relationships

Once we produce our first cross-tab, we can look at it as just a summary of our list of observations and variables (remember Table 1.1?). This is what people who are not really familiar with cross-tabs typically do (or, at least, what I think those people do). Actually, that was what I myself used to do before being introduced to the wonders of categorical data analysis. However, by using some imagination, we should start thinking of cross-tabs in a different way. Let's picture this: since the two categorical variables under analysis are *cross-tabulated* against one another, we have the opportunity to use our cross-tab to formally understand how they *relate* to each other. In other words, we have the basis to verify whether there is a *relationship* between that pair of variables.

I do foresee that the relationship between two categorical variables is quite a hard concept to grasp at the beginning. I do also have the feeling that, when thinking about relationship, many people with some familiarity with statistics will think of scatterplots of two numerical variables; for example, a scatterplot of stature against body weight. We know that, if the two variables are related to one another in some way, we will see a cloud of points in the resulting scatterplot, and that cloud of points will feature a given spread and orientation. The latter would give us an idea of how the two variables behave when considered together. When one increases, the other might increase (if there is a positive correlation); or it might be possible that when one increases, the other decreases (negative correlation). Compared to this quite clear-cut and easy-to-grasp scenario, thinking about a relationship between two categorical variables might prove not as easy to grasp. But (and here we go with another promise on my part) I will try to introduce the idea in layman's terms.

DOI: 10.1201/9781032726328-2

2.1.2 Grasping the Concept of Independence

Let's have a look at Table 2.1. It represents a 2 × 2 cross-tab where the GENDER of the Titanic passengers is tabulated against SURVIVAL. This is a fictitious example, which we will be using again later on (Section 2.3.2), so keep an eye on it!

We might be tempted to think that the female group has a relation (is more associated) to the survived group simply because 117 females fall in the "survived" group, whereas only 78 males fall in that group. However, this comparison is something you may not want to do. It would be like comparing apples with bananas. What does that mean? You cannot compare 117 to 78 because those counts are coming from two different column marginal frequencies (do you recall the marginal frequencies we spoke about in Section 1.5?). In other words, yes, it is true that 117 females survived as opposed to the 78 males who survived. But those 117 females who survived come from a total of 180 females, whereas those 78 males who survived come from a total of 120 males. The larger quantity of females surviving could be a mere reflection of the overall larger number of females.

To make things comparable, we have to account for the difference in totals for females and males. We achieve that using proportions instead of raw counts. If we consider the proportion of dead and survived among males and among females, we can see that the proportion is the same for each gender. Have a look at Table 2.2.

Let's first consider the row marginal frequencies and related proportion; in other words, let's for a moment disregard GENDER and focus on the last column of the table. We have a grand total of 300 individuals, 105 of whom died and 195 survived. If we turn those counts into proportion, 35% of our passengers died (105 : 300 × 100) and 65% survived (195 : 300 × 100). Now, let's have a look at the proportion of dead and survived *within* each gender: 35% of the males (42 out of 120) died, whereas 65% of them (78 out of 120) survived. Exactly the same proportion repeats for the females: 35% of them died (63 out of 180), while 65% survived (117 out of 180). Overall, not only is the proportion of dead and survived *within* the male group and *within* the female group the same, but they are also exactly similar to the proportion of

TABLE 2.1

Cross-Tabulation of SURVIVAL Outcomes and GENDER for a Fictitious Group of 300 Titanic Passengers

	GENDER		
SURVIVAL	Male	Female	Total
Died	42	63	105
Survived	78	117	195
Total	120	180	300

TABLE 2.2

Cross-Tabulation of SURVIVAL Outcomes and GENDER for a Fictitious Group of 300 Titanic Passengers, with Percentages Based on the Total Count for Each Gender

	GENDER		
SURVIVAL	**Male**	**Female**	**Total**
Died	42	63	105
Within column (%)	35.0	35.0	35.0
Survived	78	117	195
Within column (%)	65.0	65.0	65.0
Total	120	180	300
Within column (%)	100.0	100.0	100.0

dead and survived in the overall sample (that is, ignoring GENDER). In other words, the proportion of dead and survived within each gender is a mere reflection of the proportion of dead and survived in the overall sample (keep this in mind because it will prove handy later on).

'Why on earth are you talking about percentages when you were supposed to introduce us to the idea of relationship between pairs of categorical variables?', this is what are you thinking, isn't it? Here is the answer: there is no relation between two categorical variables when whichever group you are in for one variable *does not* change the likelihood of whichever group you are in for the other variable (and vice versa).

In the context of our specific example, there would not be a relationship between GENDER and SURVIVAL if whichever group you are in for GENDER *does not* change the likelihood of whichever group you are in for SURVIVAL (and vice versa). What does that mean? Given the proportions described above and represented in Table 2.2, if you are a male travelling on the Titanic (in this particular fictitious example), you are more likely to survive. If I randomly draw one male out of those 120, there is a 65% probability that I will draw a lad who survived. And the same holds exactly true for females: if I randomly draw one female out of those 180, there is a 65% probability that I will draw a lady who survived. The same probability repeats if I ignore GENDER: if I were to pick up a random passenger out of the 300, there would be a 65% chance that I would pick a passenger who survived, exactly as it would happen when not ignoring GENDER.

2.1.3 Wrapping up on Independence

In our fictitious example, whichever group of the GENDER variable I am in *does not* change the probability of whichever group I am in of the SURVIVAL variable. Males are as likely to survive as females, and the probability for each individual gender reflects the probability in the overall sample. Spoiler alert here: in reality, as we will observe later on, things went differently on the Titanic. This crafted dataset was meant to show an example of a cross-tab where *there is indeed no relation* between two categorical variables. To use more scientific terminology, the absence of a relation between pairs of categorical variables is called *independence*: in our fictitious example, GENDER and SURVIVAL *are* independent.

Well, here we are. Let's put the idea of independence to work and have some fun exploring the real Titanic data. Have a look at Table 2.3. The question here is still: are GENDER and SURVIVAL independent? If there was independence, we would expect the total number of males to be split between the "died" and "survived" groups according to the overall proportion (61.8% vs. 38.2%), and the same would hold true for the female group (this is what I asked you to keep in mind, remember?). However, it is apparent that there is a discrepancy between those proportions. More males fall in the "died" group (80.9% vs. an overall proportion of 61.8%), whereas more females fall in the "survived" group (72.7% vs. an overall proportion of 38.2%). This is a good hint at the *possible* existence of a *dependence* between the two categorical variables.

Do you recall what we previously said about independence? Let me repeat: there is *independence* (that is, no relationship) between GENDER and SURVIVAL if whichever group you are in for GENDER *does not change* the likelihood of which group you are in for SURVIVAL (and vice versa). However, as seen in Table 2.3, the proportions indicate that if I randomly draw a female out of the 466 females travelling on the Titanic, it would be more likely to draw a female who survived (about 73% probability). Instead, if I were to do the same for those 843 males, chances are that I would pick up a male who (unfortunately) died (about 81% probability). Therefore, falling in either of the two GENDER groups *does affect* the likelihood of which of the SURVIVAL groups one falls into. There seems to be *dependence* (relationship)

TABLE 2.3

Cross-Tabulation of SURVIVAL Outcomes and GENDER for the 1309 Titanic Passengers, with Percentages Indicating the Proportion of Each Survival Outcome within the Respective Gender

	GENDER		
SURVIVAL	Male	Female	Total
Died	682	127	809
Within column (%)	80.9	27.3	61.8
Survived	161	339	500
Within column (%)	19.1	72.7	38.2
Total	843	466	1309
Within column (%)	100.0	100.0	100.0

between those two variables in the real Titanic data. Or, to put it another way, the *hypothesis of independence* (that is, hypothesising that there is *no* relationship; bear this in mind too because it will prove handy soon) does not seem to hold in this specific example.

However, while proportions can give us a rough preliminary idea of what is going on in our cross-tab, we do need a formal test to be confident that what we observe in our dataset is not just *a matter of chance* (more on this soon).

2.2 Assessing Independence: The Chi-Squared Test

At the end of the preceding section, I left you with a pretty obscure sentence: 'to be sufficiently confident that what we observe in our dataset is not just a matter of chance'. Apologies for that. I do read your mind here; you must be wondering 'why should I carry out a test to verify what I actually see in the data? Data are there, in our table, right before our eyes; so, let's calculate some proportions and let's take it from there! And, by the way, why are you talking about chance?'.

That is what you are thinking, am I right? I do believe that I should clarify why we need a formal test to ascertain whether the *hypothesis of independence* (remember? I asked you to keep this in mind towards the end of the previous paragraph) holds true for any given cross-tab we aim to analyse. But, for the time being, take my word for it and hold on tight. Later on, I will explain why we need a test and what *matter of chance* means.

Let's take a deep breath and start our fascinating trip into the world of what is called the *chi-squared test*. We can use it to formally understand if the *hypothesis of independence* applies to our cross-tab. In other words, and sticking to our Titanic dataset, the chi-squared test will help us answer the question: are GENDER and SURVIVAL independent of one another? Or, to put it another way, does the hypothesis of independence hold for those two variables? The chi-squared test allows us to *test a hypothesis*, which in the case of this particular test is about the independence between a pair of categorical variables. It is not by chance that this test (as well as other statistical tests) belongs to the realm of what is called *hypothesis testing*. Having said that, we are ready to familiarise ourselves with our chi-squared test (oh dear, how many times have I repeated the word 'test' so far!). Let me just introduce an idea about its underlying logic, and then I will provide you with a few further details.

What the test does behind the scenes is to come up with a measure (for the time being let's call this a *number*, even though it should be called a *test statistic*) that summarises the amount of difference between the counts that we observe in our cross-tab and the counts that we *would see* if the two categorical variables being studied *were* independent. In other words, when we come up with a cross-tab, we do not know beforehand if the two variables are independent or not. What the chi-squared test does for us is (1) to work out the counts that we would expect in case the variables *were* independent, (2) to come up with a *number* that summarises how much the observed counts and the expected ones differ (I said this already, but repeating things may help), and (3) to establish how likely it is that such amount of difference is a just *matter of chance*. Ops, I did it again; I used this obscure expression, but, as promised, I will clarify it soon.

Enough theory: let's delve straight into an example and have a look at Table 2.4. It represents the same data shown in Table 2.3 (real Titanic data), but I added a couple of extra rows. Those two extra rows are reporting what are called *expected counts*. Those counts are the ones we would expect if the hypothesis of independence held true, that is, if GENDER and SURVIVAL were independent of one another. Think back to what we said earlier about Table 2.2: if the two variables were independent, we would expect the proportion of dead and survived within each gender to be the same (or very close to) as the overall proportion. So, if GENDER and SURVIVAL were independent, we would expect those 843 males to be split between "dead" and "survived" according to the overall proportion (the column on the right-hand side of Table 2.4). 61.8% of those 843 males (=521) would fall in the "died" group, whereas 38.2% of those 843 (=322) would fall in the "survived" group. The same would apply to the females: 61.8% of those 466 females (=288) would fall in the "died" group, whereas 38.2% of those 466 (=178) would fall in the "survived" group.

Are you still with me? Those expected counts are the counts we would expect under the hypothesis of independence, that is, hypothesising that GENDER and SURVIVAL are independent. If you are curious about how you

TABLE 2.4

Cross-Tabulation of SURVIVAL Outcomes and GENDER for the 1309 Titanic Passengers

SURVIVAL	GENDER		Total
	Male	Female	
Died			
Observed	682	127	809 (61.8%)
Expected	521	288	
Survived			
Observed	161	339	500 (38.2%)
Expected	322	178	
Total	843	466	1309 (100%)

Note: Counts expected under the hypothesis of independence are also reported. Percentages represent the proportion of each survival outcome in the total sample.

can quickly work out those expected counts, consider that they correspond to the product of the marginal totals of each level, divided by the table's grand total. For example, for the cell corresponding to males who died, the expected count is $(809 \times 843) : 1309 = 521$. By the same token, for the cell corresponding to females who survived, we have $(500 \times 466) : 1309 = 178$. Cool, eh?

So far, we have come up with two things: on the one hand, the counts that we *observe* in our cross-tab; on the other hand, the counts we would *expect* if GENDER and SURVIVAL were independent. It makes intuitive sense to think that, if the observed counts are equal or very close to the expected ones, what we observe would be very compatible with the hypothesis of independence. The more the observed counts diverge from the expected ones, the less we would trust the hypothesis of independence. The larger the divergence, the larger would our distrust in the hypothesis of independence be. Does that make sense to you? I hope so. But I do foresee your next question: how large does this divergence have to be to not trust the hypothesis of independence and to start thinking that there is indeed a dependence between GENDER and SURVIVAL? That is a very good question; believe me. But I have to postpone the answer for a while; so, again, bear with me and enjoy the ride.

As said, the test distils down the difference between the observed and the expected counts into a single number, which (as said) is generically called *test statistic*. This number is called (not very surprisingly) *chi-squared value* and is calculated according to the following formula:

$$\chi^2 = \sum_{i=1}^{n} \frac{(O_i - E_i)^2}{E_i}$$

Ok, granted; the formula looks a bit intimidating, but it is really easy to crack. It just tells us that, for every cell of our cross-tabs, we have to subtract the expected count (E) from the observed count (O), square the result (oh my goodness!), and divide the resulting number by the expected count. If you do it for every cell and then sum all the numbers you get, you will have the chi-squared value. Are you with me here? If you are not, please go back a few pages, have a look at Table 2.4, and then the following formula will be easy to decode:

$$\chi^2 = \frac{(682 - 521)^2}{521} + \frac{(127 - 288)^2}{288} + \frac{(161 - 322)^2}{322} + \frac{(339 - 178)^2}{178}$$

If you do the math, you will have:

$$\chi^2 = 49.75 + 90.00 + 80.50 + 145.63 = 365.89$$

Here it is. For the cross-tab in Table 2.4, the associated chi-squared value is 365.89. Pretty straightforward, eh?

I bet you are wondering: (1) why do we square the differences between the observed and the expected counts, and (2) why do we have to divide by the expected count? If the differences were not squared, we would have ended up having positive numbers and negative numbers that cancelled one another out, giving us 0 as a result. You can try it yourself. Essentially, we squared those differences in order for the numerators to be all positive numbers and to not end up with a chi-squared value of 0. As for the second question, by dividing the square of each difference by the expected count, we ensure that our statistic takes into account the relative size of the discrepancy, not just the absolute size.

What does it mean? Let's consider the first and the fourth terms in the above equation for the calculation of the chi-squared statistic. If we do the simple math, we realise that both terms feature a difference of 161 between the observed count and the expected count. In fact, $682 - 521 = 161$ and $339 - 178 = 161$. Do not be confused here: for the time being, I am disregarding the squaring of the differences because I have already explained why that step is needed. So, both terms (which correspond to two cells of our cross-tab) feature the same raw difference of 161. But what if we consider this difference in relative terms, that is, compared to the expected counts? A difference of 161 when we expect 521 is less pronounced than a difference of 161 when we expect 178. In the first case, the difference corresponds to $161 : 521 = 0.31$, that is, a 31% discrepancy, whereas, in the second instance, it corresponds to $161 : 178 = 0.91$, a 91% discrepancy.

Building on this rationale, dividing the squared differences by the expected counts ensures that cells featuring a larger discrepancy will contribute more to the final chi-squared statistic. In fact, the chi-squared value for the fourth term is 145.63, whereas for the first term is 49.75. Consider the division step like a fairness adjustment which ensures that each category's contribution to the final chi-squared value is proportional to the expected counts. This way, the chi-squared statistic does not just look at the raw differences; it considers how much those differences matter based on what is expected under the null hypothesis of independence. Granted; the squaring of the differences brings our chi-squared statistic to a scale that is not akin to the original scale of our observed counts. However, this is a necessary step that cannot be avoided. Later on, in Section 4.1.2, I will return to this squaring issue, even though in a slightly different context.

Back to our observed chi-squared value. The four values we previously got (49.75, 90.00, 80.50, 145.63) quantify the amount of difference between the observed and the expected count in each cross-tab's cell. The chi-squared value 365.89, which is the sum of those numbers, summarises the *overall* amount of departure in our cross-tab between the observed counts and the counts that would be expected if the hypothesis of independence applied to our data.

As I touched upon earlier on, if the observed counts and the expected ones were the same, or almost the same, it is easy to understand that that overall summary (the chi-squared value) would be a very small number. It would be 0 if the observed and the expected counts were exactly the same. Here we do not have a small number: 365.89 is not that small after all. Now the question is: after we have quantified how far apart our data are from what we would expect in case the variables were independent, how can we use the chi-squared value to ascertain how likely it is that the hypothesis of independence applies to our data? Or, to put it another way, how confident can we be in distrusting the hypothesis of independence?

In an attempt to address those questions, we are faced with some bad news and some good news. What do you want to hear first?

Let's start with the bad news. The chi-squared value in its own right does not tell us anything special. Yes, it is true: I said earlier on that it provides us with an overall summary of the amount of divergence between the observed and the expected counts. But I also said: 'how large does this divergence have to be to not trust that hypothesis of independence and to start thinking that there is indeed a dependence between GENDER and SURVIVAL?', remember? We cannot use the chi-squared value to say 'ok ... 365.89 is very large ... let's distrust the hypothesis of independence and conclude that there is indeed a dependence'. Sorry, we cannot do that. For one thing, because large, small, or medium, are relative terms (unless you are buying a pair of trousers!) that entail a comparison with some kind of yardstick. So, it makes (or should make) intuitive sense that we need to compare and contrast that 365.89 against something in order to understand if that 365.89 indicates a divergence large enough to distrust the hypothesis of independence.

Here we are: we have finally arrived at a crucial point when it comes to understanding the logic of the chi-squared test, as well as of any other statistical test in general. Every hypothesis test produces a value (remember, we called this a *test statistic*) and an associated *p-value*. The chi-squared test is no exception. Armed with this new concept (the *p*-value associated with our chi-squared statistic), which for the time being remains (I know) a sort of black box, let's move on to the next section.

2.3 Understanding Statistical Significance

2.3.1 Introduction

To shed some light on that black box, I need to take a small detour, and I will ask you to think about the difference between a *population* and a *sample*. Yes, there are cases where free food samples are given away and a large population gathers to eat for free. But that is not what we are talking about. In statistics, a *population* is the universe of all the possible subjects we may wish to study: for instance, all the teenagers living in a big city (or in a state, or in a country), or all the ants living under the topsoil in a tropical forest (or under the wooden floor of my flat, damn it!).

Typically, for many different reasons involving logistics, time, funds availability, and other factors, we are not in a position to collect data about each individual member of a given population. We typically draw (using different strategies that I am not going to cover here because I do care about your well-being) a random *sample* which, compared to an entire population, is smaller, therefore more manageable, and thus easier to study. While we focus on a sample because it is easier to be studied compared to an entire population, our interest does not really lie in the sample in its own right. We use the sample as a means towards an end. The end (our analytical goal) is to *guess* a given property of the parent population (which we *do not directly observe*) on the basis of the corresponding property that we *do observe* in our sample. This process is technically called *inference*.

I can picture you with a gigantic question mark hovering above your head: 'what's the connection between sample(s), population, and the analysis of our cross-tabs?'. Is that what you are thinking, am I right? The answer is: because when we analyse a cross-tab and we want to assess if there is dependence between the two variables of interest in our sample, what we really want to know is whether the hypothesis of independence (if any) holds in the population from which we drew our sample. As said, we use what we observe in our sample to guess (infer) what we cannot directly observe in our population because we do not have access to all its members. Is that reasonably clear? No worries if it is not. I myself was puzzled when I started reading about statistics some years ago. Let me expand on the concept and let's picture what follows, resuming our discussion about the Titanic data.

2.3.2 The Chi-Squared Test *p*-Value

Let's imagine that a detective wants to ascertain whether on the Titanic the rule whereby women are to leave the sinking ship first has been adhered to. Let's assume further that they have been given an old inventory, consisting of hundreds of dusty pages, and have a limited amount of time to build the list of passengers, taking note of each one's GENDER and SURVIVAL (remember Table 1.1?). So, to get the job done (and not be fired by the employer), while the entire official record of passengers is the *population* of interest, the detective needs to draw a small *sample* that is easier to extract and study given the time allocated.

Now, for a moment, let's assume that Table 2.2 (which we saw in Section 2.1.2) is the entire population of Titanic passengers that our detective wants to sample from. And let's assume further that we know beforehand (but we will not tell the detective; sometimes it is funny to be mean!) that in that population GENDER and SURVIVAL are *indeed* independent. For the sake of argument, I want to simulate a population where we *do* know that GENDER and SURVIVAL *are* independent, like in Table 2.2.

Please, do not be misled here: we are working with a fictitious scenario; those 300 passengers in Table 2.2 represent the *entire* Titanic *population* which the time-pressured detective is going to sample from. And, as said, in this population the two variables *are* independent. Our detective draws a random sample of 100 passengers from that population of 300, and they come up with Table 2.5.

The chi-squared value for that table is very small (0.160; take my word for it, I made the calculation at home), indicating that the amount of departure between the observed and the expected counts is tiny. Notice that here, to keep things simple, I am not reporting the expected counts. What I am reporting is the proportion of dead and survived passengers for the entire sample (last column to the right), and the proportion of dead and survived individuals within each gender. Not surprisingly, like in the parent population, males are about as likely as females to fall in the survived group. This *child* random sample is pretty close to the *parent* population.

Now, let's rewind the tape of the detective's life, and watch them drawing another random sample. Have a look at Table 2.6.

Here, the chi-squared value is comparatively larger (0.447), indicating a larger departure between the observed and expected counts than in the previous random sample. We have the impression that there is a difference between males and females. Females are comparatively more likely than males to fall in the "survived" group (65% vs. 58%), whereas males are more likely than females to fall in the "died" group (42% vs. 35%). Also, there is a larger-than-average proportion of females in the "survived" groups (65% vs. an average of 62%), while there is a larger-than-average proportion of males in the "died" group (42% vs. an average of 38%). Even from this standpoint, it would seem that females and males are more likely to fall in two different groups of the SURVIVAL variable. This time round the child sample proves a bit different from the parent population. You know, it may well randomly

TABLE 2.5

Cross-Tabulation of SURVIVAL Outcomes and GENDER for a Fictitious Sample of 100 Titanic Passengers Randomly Drawn from the Data in Table 2.2

	GENDER		
SURVIVAL	Male	Female	Total
Died	12	22	34
Within column (%)	32.0	35.0	34.0
Survived	26	40	66
Within column (%)	68.0	65.0	66.0
Total	38	62	100
Within column (%)	100.0	100.0	100.0

Note: Percentages based on the total count for each gender.

TABLE 2.6

Cross-tabulation of SURVIVAL Outcomes and GENDER for a
New Fictitious Sample of 100 Titanic Passengers Randomly
Drawn from the Data in Table 2.2

SURVIVAL	GENDER		
	Male	Female	Total
Died	18	20	38
Within column (%)	41.9	35.1	38.0
Survived	25	37	62
Within column (%)	58.1	64.9	62.0
Total	43	57	100
Within column (%)	100.0	100.0	100.0

Note: Percentages based on the total count for each gender.

happen that one child (out of a number of children) is naughty even when
parents are well-behaved!

These are just two examples of how variable a cross-tab and the associated chi-squared value would be if the time-pressured detective would keep
drawing random samples from those 300 passengers that (let me stress it
again) we are considering *the* entire population available where we *do* know
(but the detective does not) that GENDER and SURVIVAL *are* independent.
All the cross-tabs that we could repeatedly generate along this imaginary
process would never look always alike, and the associated chi-squared value
would not always be exactly the same across all the random samples. There
would always be some sort of *random variability*. Along the imaginary process, there would always be a number of naughty random children cross-
tabs that would give the impression of some dependence between GENDER
and SURVIVAL, even though in the parent population those two variables
are indeed independent.

2.3.3 Independence and the Variability of the Chi-Squared Statistic

If the detective would repeat the process many times (say, 1000), and
every time they would keep track of the chi-squared value associated
with each cross-tab of 100 random passengers, they could produce the
frequency distribution histogram of the chi-squared values illustrated
in Figure 2.1.

In this frequency distribution histogram, the horizontal axis represents
chi-squared values, while the height of each bar (histogram) indicates how
many chi-squared values fall in the interval represented by that bar. For
example, the first bar to the left represents chi-squared values between 0 and
0.5, and 746 (out of 1000, or about 75%) chi-squared values fall in this interval

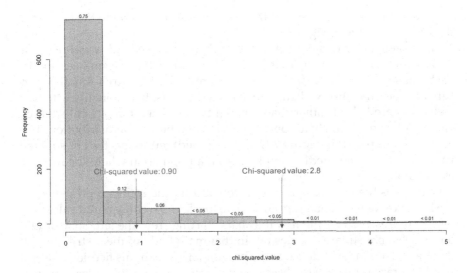

FIGURE 2.1
Frequency distribution histogram of the values of the chi-squared statistic calculated by randomly drawing 1000 samples of 100 observations from a fictitious population (like the one represented in Table 2.2) where there is independence between two categorical variables (each one featuring two levels). The horizontal axis represents the chi-squared value; the height of each bar indicates how many chi-squared values fall in the interval represented by each bar. The proportion of chi-squared values (out of 1000) falling in each interval is indicated above each bar. The plot represents the distribution of the chi-squared statistic under the hypothesis that the two variables are independent in the population (null hypothesis). A chi-squared value equal to or larger than (say) 2.8 would fall in the slim right tail of the distribution and would be observed 5% of the time or less. Random samples producing smaller chi-squared values (say 0.90) would fall in the main body of the distribution and would be far more likely to be observed under null hypothesis of independence.

(this latter piece of information is visually represented by the height of the bar itself).

What does Figure 2.1 teach us? When there is *independence* in the population (as in our fictitious case of 300 passengers from which the detective has been drawing samples), the majority of the chi-squared values would be tiny (because the majority of the randomly drawn samples are well-mannered children). In the simulation that I ran to write this section, 746 out of 1000 (0.746 = 75%) were between 0 and 0.5. Many of the remaining were still small in value (indicating a small departure between the observed and expected counts) and were *less and less* frequent (see how the bars become shorter and shorter as the value of the chi-squared statistic increases). In *fewer and fewer* cases, the random samples produced comparatively larger chi-squared values, indicating a comparatively larger departure between the observed and the expected counts. But these (naughty children) are pretty rare: 5 out of 1000 (0.005, so way less than 5%) produced a chi-squared value between 3 and 3.5; 2 out of 1000 (0.002, again way less than 5%) produced a chi-squared value between 4 and 4.5 (look at the extremely short second to the last bar to the right-hand side of Figure 2.1).

What seen so far demonstrates that, if there is indeed *independence* in the population, it is more likely that *the* sample we draw (I gave emphasis to that *the* because we typically draw only *one* sample) will feature a small departure between the observed and the expected counts. It is more likely to be a well-mannered child rather than a naughty one. Therefore, its chi-squared value will be more likely to come from the main body of a distribution such as the one portrayed in Figure 2.1, which (as said) represents the chi-squared values obtained repeatedly sampling from a population where the hypothesis of independence *does* hold.

What does the *main body* mean? If you add up the counts represented by the bars, you will find out that 951 (746 + 116 + 55 + 34) chi-squared values are overall represented by the first four bars starting from the left. So, 95% (951 out of 1000) chi-squared values fall in that main body of the distribution. If there is indeed independence in the population (like in this fictitious case), it would be rare (unlikely) to observe comparatively large chi-squared values, like the ones falling in the right tiny tail of our distribution. Such large chi-squared values (indicating a larger departure between the observed and the expected values) would be rare/unlikely when there is indeed independence in the population.

Let's come back to our detective, and picture this. They cannot rewind the tape of their life like we did and they cannot repeatedly draw random samples from the population of 300 passengers. What the detective can do is drawing one random sample and carrying out the chi-squared test. If GENDER and SURVIVAL are independent in the population, getting a large chi-squared value just as a matter of *chance* (*random variability*) would be unlikely. It would happen 5% of the times or less. If GENDER and SURVIVAL are independent in the population, getting a chi-squared value that falls in the main body of our theoretical distribution would be more likely to happen; it would happen 95% of the time on average.

Now, let's have a look at some other details of Figure 2.1. A chi-squared value *equal to or larger* than (say) 2.8 would fall in the slim right tail of the distribution, and would be observed 5% *of the time or less*. So, all in all, it would not be very likely that, sampling from a population where the two variables *are* independent, we would observe a chi-squared value of such size (or larger). Random samples producing smaller chi-squared values (like 0.90 in Figure 2.1, pointing to a tiny difference between the observed and expected counts) would be far more likely to be drawn.

Before moving on to the next section, I want to ask you whether you noticed that we can associate any chi-squared value with a probability score expressing how likely it would be observing that value under the hypothesis of independence? Keep this in mind because (as you know by now) things can soon turn out to be pretty useful.

2.3.4 *p*-Value and the 0.05 Threshold

After all this long detour, let's get back to the earlier question: how large the divergence between observed and expected counts (as summarised by the chi-squared statistic) has to be in order for us to distrust the hypothesis of independence and to start thinking that there is indeed a dependence between GENDER and SURVIVAL? We are now in a position to address that question. Since we cannot use the chi-squared value in its own right, we have to set it against the distribution that portrays how the chi-squared statistic is distributed under the hypothesis of independence. As said, if the observed chi-squared statistic falls in the tail of that theoretical distribution, it means that it would be rarely observed if GENDER and SURVIVAL were independent in the population from which we drew our sample.

At this point, I will omit some intricate details. It suffices for you to know a couple of things. The distribution of the chi-squared statistic under the null hypothesis of independence can be theoretically (but also empirically; more on this in Section 2.3.5) derived by knowing a key ingredient that is called *degrees of freedom* (hereafter *df*). To admittedly over-simplify a complex definition, we can conceive the df as a number that quantifies how large is a table; it is equal to the number of rows minus one multiplied by the number of columns minus one. Once we know our cross-tab's df, the distribution of the chi-squared statistic under the hypothesis of independence can be theoretically worked out, and the observed chi-squared value can be assessed against it in order to work out its associated probability. If it turns out that the chi-squared value falls somewhere in the tail of the distribution (so its associated probability is low), it means that it would be unlikely if the variables were independent in the parent population. It is worth noting that when our table has more than two levels in either or both of the two variables being analysed (that is, when the cross-tab is larger than 2 × 2), the df increase, causing the chi-squared distribution to transform. Specifically, as the df get larger, the distribution progressively assumes a more symmetric and bell-shaped profile (different from the one in Figure 2.1), analogous to the tapering shape of a party hat. Think of it as the chi-squared's way of adjusting its hat for a bigger party.

How unlikely does the observed chi-squared value need to be to qualify as unlikely? How small does the associated probability need to be to be considered small? Statisticians have established a threshold of 0.05 (or 5%). If we would observe the chi-squared value 5% of the time or less (that is, it has a probability ≤ 5%), then we can reasonably doubt the hypothesis of independence. This means we can be 95% (or more) confident that GENDER and SURVIVAL are dependent in the parent population. The number that quantifies the probability of observing the obtained chi-squared statistic under the hypothesis of independence is called *p-value*. If the chi-squared statistic is associated with a *p*-value less than 0.05, the test result is deemed *statistically significant*. In other words, the test indicates that there is a statistically significant dependence between our two variables in the parent population.

To wrap up: if the previously discussed frequency distribution represents how variable the chi-squared statistic can be due to random variability when the two variables *are* indeed independent in the parent population, a small *p-value* (≤ 0.05) associated with the observed chi-squared statistic indicates that our chi-squared value would be a rare and unlikely result of random variability *if* the two variables *were independent* in the parent population. We have some objective basis to distrust the hypothesis that the variables are independent in the parent population. Wow ... we made it!

Building on our theoretical foundation, let's now apply these concepts to the Titanic dataset we presented in Table 2.4. The chi-squared statistic is 365.89, the df of the table are 1 (that is, $2 - 1 \times 2 - 1$), and the associated *p*-value is smaller than 0.05 (actually, it is even smaller than 0.001). The test proves statistically significant. What does all this mean, and what conclusions we would arrive at on the basis of the test?

It tells us that if the hypothesis of independence were true (that is, under the hypothesis that the two variables *are not* associated in the population), it would be *unlikely* to observe a chi-squared value of that size (or larger). The probability of observing it would be way smaller than 5% (and even smaller than 1%). We can distrust (or, more technically, *reject*) the hypothesis of independence (aka *Null Hypothesis*) and we can conclude that GENDER and SURVIVAL are likely to be associated in the population. The dependence that we observe in our sample is *not likely* to be just the result of *random variability*; rather, it is likely to reflect a dependence between the two variables under study that exists in the population from which we drew our sample.

Finally (and I do promise we are almost done with this section), think of the *p*-value as a measure of how consistent our cross-tabulated data are with the hypothesis of independence. For instance, a *p*-value of 0.05 (=5%) suggests that if the hypothesis of independence is actually true, there is a 5% chance of observing a dataset with a level of deviation from this hypothesis as large as ours, or more, due to random variation. Similarly, a *p*-value of 0.01 (=1%) means there is a 1% chance of observing such a deviation under the same assumption. When the *p*-value is very low, it suggests that the observed data do not align well with what we would expect if the hypothesis of independence were true. Thus, we consider rejecting the null hypothesis. If our cross-tabulation strongly suggests a mismatch with the hypothesis of independence, it leads us to conclude that it is more likely that an association exists between the two variables in the parent population.

Consider an analogy: suppose the weather forecast indicates a 0.05 probability of rain tomorrow. You might reject the null hypothesis of *staying home* in favour of the alternative hypothesis of *going for a picnic*, reasoning that it is quite unlikely that you and your relatives would get drenched. There is still a possibility of rain, but it is slim. In this analogy, choosing to stay home is akin to accepting the null hypothesis of independence between two variables in the parent population. Conversely, deciding to go for a picnic represents the conclusion that these variables are likely dependent in the parent population.

2.3.5 The Bottom-Up Approach to Significance: Monte Carlo Simulated *p*-Value

Having explored the concept of the *p*-value in relation to the chi-squared statistic and its significance threshold, let's introduce a complementary approach that can enhance our understanding of those concepts: the Monte Carlo method. This approach may not only simplify our grasp of the distribution of the test statistic under the null hypothesis, but can also prove useful in situations where the chi-squared test faces limitations, for instance, in cases involving small expected counts (see Sections 3.3 and 3.4).

Named after the famous Monte Carlo Casino, known for its random outcomes, the Monte Carlo method allows for empirical observation and understanding of the chi-squared statistic's variability and distribution. This is achieved by generating a multitude of simulated datasets under the null hypothesis of independence. This approach not only simplifies our understanding of the distribution of the test statistic under the null hypothesis but also bridges the gap between theoretical probability and practical application. It represents a shift towards a more empirical data-driven approach, beneficial both pedagogically, for offering a practical representation of the statistical concepts, and from the standpoint of actual cross-tab analysis. Let's delve deeper into this approach.

The Monte Carlo method allows to compute the chi-squared test's *p*-value through simulations. This procedure mirrors the detective's repeated sampling from Section 2.3.3 in a structured manner. Under the assumption that GENDER and SURVIVAL are independent, this method generates a range of alternative datasets under that premise. This process, a kind of *replaying* or *rewinding*, enables us to empirically create a distribution of chi-squared values. This distribution represents what we might expect to see by chance alone, in the absence of any actual association between the variables in the population. The generated distribution differs from a theoretical one as it is based directly on our data. It is created empirically, not derived from theory.

Suppose our observed chi-squared value is 30. Now, imagine we draw 1000 new samples, generated under the assumption of no relationship between GENDER and SURVIVAL, and we calculate a chi-squared value for each. Let's say 25 of these simulated datasets produce a chi-squared value equal to or greater than 30. The Monte Carlo *p*-value, computed as $25 : 1000 = 0.025$ or 2.5%, indicates the proportion of times a chi-squared value of 30 or more might occur, purely by chance, under our initial assumption of independence. As you well know by now, it would be considered statistically significant.

To demonstrate the Monte Carlo method in our Titanic example (Table 2.4), we opt to generate 10,000 random cross-tabs that have the same row and column sums as the observed table. In other words, the counts in the individual cells can vary across the random tables, as long as the row and column marginal sums are the same as in the cross-tab representing our observed data. As for the number of random tables to generate, you might wonder why 10,000 and not 100 or a million. The number of simulations you run also influences the smallest p-value you can possibly observe. Specifically, the smallest p-value we can get is calculated as 1 divided by 1 plus the number of simulations, or $1 : (1 + 10,000)$ in our case. This means the smallest p-value we can possibly get here is 0.0000999. More simulations mean a finer resolution of the p-value but at the cost of increased computational demands, presenting a trade-off between precision and practicality.

For the sake of argument, since we do not (and never will) have enough space to list 10,000 tables, we limit ourselves to having a sneak peek at four of them. Let's have a look at Table 2.7. As you can see, the individual counts randomly vary across tables, but the row and column totals remain exactly the same. Furthermore, they correspond to the row and column totals of the original data (Table 2.4). We compute the chi-squared statistic for each random cross-tab. From Table 2.7(A–D), the chi-squared value is 0.691, 0.014, 0.226, and 0.353, respectively. All of them are not far from 0, indicating a small departure from the hypothesis of independence. One of them, namely

TABLE 2.7

(A–D) Monte Carlo Simulated Cross-Tabulations Based on the Marginal Totals from Table 2.4

A	GENDER			B	GENDER		
SURVIVAL	Male	Female	Total	SURVIVAL	Male	Female	Total
Died	514	295	809	Died	520	289	809
Survived	329	171	500	Survived	323	177	500
Total	843	466	1309	Total	843	466	1309
C	GENDER			D	GENDER		
SURVIVAL	Male	Female	Total	SURVIVAL	Male	Female	Total
Died	517	292	809	Died	526	283	809
Survived	326	174	500	Survived	317	183	500
Total	843	466	1309	Total	843	466	1309

Note: These four tables showcase the variability of the chi-squared statistic under the null hypothesis of independence. Each simulation adheres to the same marginal totals for GENDER and SURVIVAL outcomes as in the observed data of the 1309 Titanic passengers (Table 2.4). Chi-squared values: 0.691 (A), 0.014 (B), 0.226 (C), and 0.353 (D).

the one yielded by Table 2.7B, is the closest to 0 by comparison, meaning that the counts in that random cross-tab are extremely close to the expected ones. In fact, for instance, the expected count for males who died is (809×843) : $1309 = 521$ and for females who survived is $(500 \times 466) : 1309 = 178$. The other expected counts are very close to the observed ones as well; you can do the simple math to check that.

These four tables, and the other 9996 that I have not reproduced here, give us an idea of how much variable the chi-squared statistic can be when the null hypothesis of no association (independence) is actually true. In fact, the random cross-tabs have been generated by crafting new tables that (as said) strictly maintain the original marginal totals of GENDER and SURVIVAL. This process is akin to creating numerous possible versions of our data where the overall numbers of males, females, dead, and survivors match those of the actual Titanic data, but are distributed differently within each random table. This method is designed to simulate what the distribution of pas-sengers might look like under numerous random circumstances, all while adhering to the assumption that GENDER and SURVIVAL are independent. By doing so, we generate a broad spectrum of outcomes that provide an empirical basis to estimate the likelihood of our observed chi-squared statis-tic arising just by chance.

Once we calculate the chi-squared statistic for each of the 10,000 random tables, we count the instances where the statistic is equal to or exceeds our observed value, which (as seen in Section 2.2) was 365.89. The test I carried out when writing this section produced a p-value equal to 0.000099. This implies that about 1 random table out of the 10,000 generated under the hypothesis of independence yielded, just by random chance, a chi-squared value equal to or larger than 365.89. The extremely low simulated p-value suggests that the observed chi-squared statistic is highly unlikely to have occurred by chance alone if GENDER and SURVIVAL were not associated, casting doubt on the hypothesis of independence. Therefore, we have bases to conclude that the two variables are likely to be associated because the probability that our data are compatible with the null hypothesis is tiny indeed.

One thing to bear in mind is that the Monte Carlo method, being rooted in randomisation, will produce subtle differences in the p-value with each series of simulations conducted. Because of the randomised nature of the process, running it a number of times might result in slightly varied results from one set of simulations to the next. Nevertheless, such small fluctuations should not significantly affect the general conclusions drawn about the sta-tistical significance of the observed association.

In conclusion, through the Monte Carlo simulation, we gain not only a deeper statistical insight but also a more tangible grasp of how rare or com-mon our observed chi-squared statistic under the null hypothesis can be, providing us with a clearer understanding of the significance of the relation-ship between the variables under analysis.

2.3.6 Beyond the Titanic: Conceptualising Statistical Generalisation

In the final reflection of this chapter, I would like to draw your attention to the broader context of our analysis of the Titanic dataset. The example I have used, involving a detective sampling from the Titanic's official record, served as a metaphor for understanding statistical significance and the principles of sampling from a larger population. However, this example, while illustrative, leads us to ponder a more profound question: what is the broader population to which we might generalise our findings from the actual Titanic data? Questions of this type are crucial in statistics, as they challenge us to consider the applicability and relevance of our findings beyond the immediate data.

One might posit that the sample of 1309 passengers, which we have analysed in this chapter, represents not merely those who travelled on the Titanic but a subset of a larger hypothetical population. This larger group could be envisioned as all potential passengers from different embarkation ports who might have chosen to journey on the Titanic on that unlucky voyage. In this context, our sample transcends its historical confines, becoming a representation of a broader population of early 20th-century transatlantic travellers.

Such an extrapolation allows us to view the Titanic not just as an isolated incident but as a window into the travel dynamics and social strata of that era. This perspective aligns with our statistical goal that, as I touched upon earlier on (Section 2.3.1), is to draw inferences about populations from samples. By considering the Titanic passengers as a subset of a larger theoretical group, we gain insights not only into the tragic event itself but also into the broader sociodemographic patterns of the time. Thus, the statistical significance we derive from this dataset is not confined to the specifics of the Titanic but extends to understanding the interplay of factors like GENDER and SURVIVAL in a wider historical context. This approach embodies the essence of statistical analysis: using information obtained from the sample we observe to shed light on a larger unobserved population.

2.4 Key Takeaways

- When analysing cross-tabs, the goal is not just understanding the sample but making inferences about the larger population.
- Random samples can vary from the population and even show relationships not present in the population. This variability is natural due to the randomness of sampling.
- It is essential to set the observed chi-squared statistic against a theoretical distribution, specifically the expected distribution of this

statistic when there is no real relationship between the variables (known as the Null Hypothesis). This comparison helps us understand whether the dependence we see in the sample is just due to chance or if it reflects a true relationship that exists in the parent population.

- Statisticians typically use a 0.05 (5%) threshold to determine statistical significance. If the observed chi-squared value would occur 5% of the time or less (that is, it has a probability $\leq 5\%$) under the null hypothesis of independence, then we reject that hypothesis. Essentially, a p-value less than 0.05 indicates there is a statistically significant relationship between the variables under study.

- The p-value tells us the probability of observing a particular chi-squared value (or more extreme) under the assumption that the two variables are independent in the population. A small p-value (<0.05) means that the observed chi-squared statistic would be a rare result of random chance if the two variables were indeed independent in the population. Thus, a small p-value gives reason to distrust or reject the hypothesis that the variables are independent in the parent population. Think of the p-value as the probability that there is no relationship (independence) between the two variables in the parent population.

- The Monte Carlo method of empirical simulation offers dual advantages: it serves as an educational tool by providing a more intuitive and data-centric understanding of p-values, and also stands as a robust analytical technique particularly when the validity of the chi-squared test is in question.

- In the Monte Carlo method as applied to the chi-squared test, numerous alternative cross-tabs are generated under the assumption of variable independence. For each of these cross-tabs, the chi-squared value is calculated. By comparing the chi-squared values from these simulated datasets to the observed chi-squared statistic, a simulated p-value is computed.

- The Monte Carlo p-value quantifies the proportion of simulated chi-squared values that are equal to or greater than the observed statistic, thereby gauging the likelihood of the observed chi-squared statistic occurring under the hypothesis of independence.

3

The Chi-Squared Test: Advanced Insights

3.1 The Smoking Gun: Tracking Down Chi-Squared Residuals

3.1.1 Standardised Residuals

As seen in Section 2.2, the chi-squared statistic is an overall measure of the amount of divergence (difference) between the observed and expected counts. It gives us a general view of the amount of departure from the hypothesis of independence. While it is true that the chi-squared statistic and its associated *p*-value tell us whether there is a *significant* association between the two categorical variables under analysis, they do not help us pinpoint *where* the significant association actually lies. If we just limit ourselves to performing the test, we would not be able to identify which observed count(s) significantly differs from the expected one(s). We need to take the analysis one step further and consider what are called Pearson's *standardised residuals* (named after the British statistician C. Pearson who is the father of the chi-squared analysis).

The raw residual in any given cross-tab's cell is the difference between the observed count and the expected count in that specific cell. However, the standardised residuals are residuals that are re-expressed in a way that provides us with some objective basis to understand whether the size of each residual can be considered so large to indicate a significant divergence from the null hypothesis of independence. No worries; I know that I need to clarify that. As I said a few pages ago, unless we buy a pair of trousers, the labels small, medium, or large are relative terms that do need some *standardised* measure in order to be agreed upon. In other words, in order for me to consider small what you considered small (or medium, or large) as well, we need to use a common unit of measure (a sort of ruler we can both agree upon).

This is what standardising a residual actually does. It transforms the raw residuals into something that we can all agree upon when it comes to judging their size. The standardised residuals are calculated using the following formula:

$$\text{standarsided residual} = \frac{O - E}{\sqrt{E}}$$

DOI: 10.1201/9781032726328-3

Again, let's crack it. For any given cell of a cross-tab, we subtract the expected count (*E*) from the observed count (*O*), and divide the result by the square root of the expected count. As simple as that. If I were to be more technical, I should say that a standardised residual is a way of re-expressing the raw residuals in terms of *z-score*. The z-score is a number that measures how distant is the value of a numerical variable in terms of standard deviations from the mean of that variable. In this case, the mean is set to 0, and the standard deviation to 1 (well ... not exactly; see Section 3.1.2). A z-score larger than 1.96 indicates that a value is considerably larger than the mean as it lies about two standard deviations above the mean. By the same token, a value associated with a z-score of −1.96 is considerably smaller than the mean as it lies at about two standard deviations below to mean of the variable. This is exactly the way the chi-squared standardised residuals should be interpreted.

To say the truth, the story is a bit more complex than that since the standardised residuals have been shown to be biased; that is, they do not do a great job at measuring the difference between the expected and the observed counts and they should be therefore *adjusted*. I will elaborate on this in Section 3.1.2. For the time being, let's have a look at Figure 3.1, which shows standardised residuals for the real Titanic dataset that was presented in Table 2.3.

How should we interpret those residuals, and what conclusion should we arrive at? For any given cell, a standardised residual larger than 1.96 (in terms of absolute value) indicates a significant dependence between the two categorical variables in that specific cell (a significant difference between the observed and the expected counts). Positive residuals derive from observed counts that are larger than expected. Negative residuals derive from observed counts that are smaller than expected. Positive and negative residuals can be conceived as indicating a sort of positive and negative association between

	Male	Female
Died	7.054	-9.487
Survived	-8.972	12.068

BLUE: significant negative residuals (< -1.96)
RED: significant positive residuals (> 1.96)

FIGURE 3.1
Chi-squared standardised residuals. They are calculated by subtracting the expected count from the observed count and dividing the result by the square root of the expected count. A residual with absolute value larger than 1.96 indicates a statistically significant departure from the hypothesis of independence. For the limitations of the standardised residuals (not being truly standardised), see Section 3.1.2. The annotation regarding the colour of the residuals value refers to the colour version of the figure. Based on the data in Table 2.3. See also Figure 3.2.

the corresponding groups of the two categorical variables under analysis. In our Titanic dataset, there is a positive (and significant) association between the males and "died" group, and a positive (and significant) association between the females and "survived" group. It is apparent that there is a significant larger-than-expected quantity of males in the "died" group (and a significant smaller-than-expected quantity in the "survived" group), while the opposite holds true for the females.

Now, let me add a couple of further details about the significance of the dependence as indicated by the standardised residuals. If you think about what we said earlier on about the p-value associated with the chi-squared statistic, it would be easy to understand that, the same way the chi-squared value is associated with a probability value (its p-value), we can associate the z-score with a probability value.

To cut a long and complex story short, if the absolute value of a standardised residual is equal to or larger than 1.96, its associated probability is at least 0.05 (that is, 0.05 or less). In other words, if the two categorical variables being analysed were independent in the parent population, the probability of observing a difference between the observed and the expected count such as the one producing a standardised residual equal to or larger than 1.96 (in terms of absolute value) would be a rare and unlikely product of random variability. It would happen 5% of the times or less if the two variables were indeed independent in the parent population.

Commenting on Figure 3.1, we said that there is a positive (and significant) association between the males and "died" group. The standardised residual is way larger than 1.96 (in terms of absolute value); it is actually 7.05. Its associated p-value is way smaller than the 0.05 threshold to which we made reference earlier on when elaborating on the p-value associated with the chi-squared statistic. Such a large positive residual would be a rare and unlikely product of random variability if GENDER and SURVIVAL were independent in the parent population. The probability of observing it as a result of pure chance would be way less than 5%.

3.1.2 Adjusted Standardised Residuals

As said in the preceding section, the chi-squared standardised residuals have been demonstrated to be biased. They tend to underestimate the significance of differences between the observed and the expected counts. It is like wanting to measure the length of a table in metres, but our measuring tape uses, as the unit of measure, a metre that is 90 cm long. We would end up underestimating how long our table really is. Without going into further stats details, it suffices for you to know that we are better off using the so-called *adjusted standardised residuals*. The adjusted standardised residuals are standardised residuals that are *adjusted* in order to take into account the row and column proportions. These types of residuals are calculated according to the following formula:

Adjusted stand.residual

$$= \frac{O-E}{\sqrt{E(1-\text{row proportion})(1-\text{column proportion})}}$$

This formula should sound pretty familiar because it is not really different from the previous one. It means that the adjusted standardised residual for a given cell is equal to the difference between the observed (O) and expected (E) count divided by the square root of the expected count, the latter multiplied by one minus the row proportion times one minus the column proportion.

Let's have a look at the adjusted standardised residuals for the Titanic dataset (Table 2.3), shown in Figure 3.2. Let me show you how we arrive at the standardised residual for the upper-left cell of Table 2.4. We can see that the observed count in that cell is 682 and the expected one is 521. Also, we can see that the proportion of that row is 0.618 (that is, there are 809 dead passengers in total, out of a table's grand total of 1309 passengers; so 809 : 1309 = 0.618 or 61.8%). Also, the proportion of the respective column is 0.64 (843 males out of a table's grand total of 1309; so 843 : 1309 = 0.64). Now, we have all the ingredients to calculate the adjusted standardised residual for that cell:

$$\frac{682-521}{\sqrt{521(1-0.618)(1-0.64)}} = 19.129$$

As simple as that! But you do not have to worry; you will not have to do the calculation by hand, since any computer stats software would do that for you. We can notice that the adjusted standardised residual is larger than the standardised one. This happens because the adjustment increases the residual if the expected count is small relative to the sample size. Anyway, even though the size of these adjusted standardised residuals is larger compared to the size of the standardised ones, the interpretation of their significance

	Male	Female
Died	19.129	-19.128
Survived	-19.128	19.129

BLUE: significant negative residuals (< -1.96)
RED: significant positive residuals (> 1.96)

FIGURE 3.2
Chi-squared adjusted standardised residuals. They are equal to the difference between the observed and expected count divided by the square root of the expected count, the latter multiplied by one minus the row proportion times one minus the column proportion. They correct a deficiency of the standardised residuals. The same interpretation provided in Figure 3.1 holds.

still holds. The cell we have taken into account features a significant association (a significant difference between the observed and the expected count)
between the "male" and the "died" group.

3.1.3 Wrapping Up on the Standardised Residuals Analysis

We can think of a standardised residual as a score that tells us whether the
observed count in a particular cell of our cross-tab is consistent with the
hypothesis of independence. If the residual is close to zero, it means that the
observed count is consistent with the hypothesis of independence and, therefore, there is nothing unusual or unexpected about it. It would be extremely
likely if the variables were independent in the parent population.

However, if a standardised residual is equal to or larger than 1.96, it means
that the observed count in that cell is much larger than we would expect if
the variables were independent (that is, under the hypothesis of independence) in the parent population. The observed count giving rise to such standardised residual (or to a larger one) would be unlikely if the hypothesis of
independence held true. By the same token, if the standardised residual is
equal to or larger than –1.96, it means that the observed count in that cell is
much smaller than we would expect if the variables were independent in the
parent population. The observed count giving rise to such a standardised
residual (or to a larger one) would be unlikely if the two variables under
study were independent in the parent population.

Now, I can again foresee your next question: 'when should we inspect our
adjusted standardised residuals?'. Luckily enough, the question is pretty
easy to answer. If you run the chi-squared test and it proves significant (associated p-value ≤ 0.05), you should take the extra step of having a look at the
adjusted standardised residuals. Why? Because it could well happen that not
all the cross-tab's cells feature a significant difference between the observed
and the expected counts. Only a few cells might *significantly contribute* to
the rejection of the hypothesis of independence. The analysis of the adjusted
standardised residuals helps us pinpoint *where* significant associations lie. If
the chi-squared test does not prove significant, there is no point in inspecting the residuals (either adjusted or not) because the overall test does not
justify the rejection of the null hypothesis of independence. In such circum-
stance, there would not be any significant dependence to pinpoint among
our cross-tab's cells.

3.1.4 From Tiles to Tales: Visualising Residuals with Mosaic Plots

If you are a person more inclined to discern patterns from visual representations, the so-called mosaic plot might be a good choice when it comes to
visualising the chi-squared residuals. Let's have a look at Figure 3.3.

To begin with, let me describe the plot in detail, dissecting every compo-
nent in turn. From a glance, it is pretty evident that it is composed of four

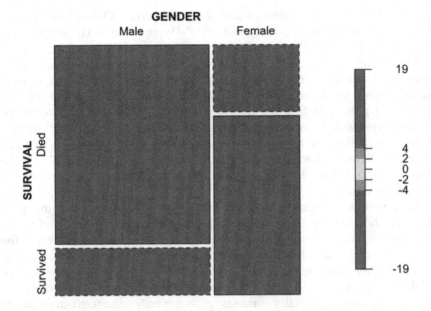

FIGURE 3.3
Mosaic plot illustrating the relationship between GENDER and SURVIVAL on the Titanic. The four tiles represent distinct groups: top-left for males who died, moving clockwise to females who died, females who survived, and males who survived. Tile height reflects survival proportions within the GENDER categories. Tile width signifies the overall gender proportion regardless of survival (64% males, 36% females). The grey-scale colour gradient reflects the adjusted standardised residuals. Tiles with a dashed border feature negative residual. A lower-than-expected number of males survived, while less females died than expected under the hypothesis of independence. Based on Table 2.3. See also Figure 3.2.

rectangular tiles, each one corresponding to one cell of our cross-tab. The top-left cell represents the males who died. Moving clockwise, the second tile represents females who died; the third cell corresponds to females who survived, and the fourth represents males who survived.

Observing the height of the tiles, some are noticeably taller while others are shorter. If there was independence in our cross-tab, the tiles would have had the same height and would have been aligned in a grid-like fashion. However, this is not the case with our Titanic dataset. Let's focus on the "male" category. The different heights of the two tiles tell us that there is a larger-than-average proportion of individuals who died among the males (682 out of 843; 80.9%). If we take into account the "female" category, we can visually appreciate that the trend is the opposite, with 339 females who survived out of a total of 466 females (amounting to 72.7%). There is a larger proportion of survivors among the females.

Before moving on to the description of the grey-scale colour featuring each tile, I will ask you a question: have you noted that the tiles feature a different width? This is because the width gives an idea of the gender proportion in the overall sample, regardless of the survival outcome. Out of 1309 passengers, 843 (64%) were males, whereas 466 (36%) were females. The width of the tiles is scaled to reflect that proportion.

What about the tile colours? They offer crucial insights into whether the observed differences in proportions are statistically significant. In fact, each tile is given a shade of grey (or a different colour, if colours are used to render the plot) reflecting the magnitude of the associated adjusted standardised residual. As you can see, a colour scale is represented on the right-hand side of the plot to signify positive and negative residuals. We can compare the table of adjusted standardised residuals in Figure 3.2 and the mosaic plot in Figure 3.3. In the latter, tiles with a dashed border feature a negative residual.

As one can see, there is a significant negative deviation from independence in the Male-Survived cell (less males survived than expected under the hypothesis of independence), and a significant negative deviation in the Female-Died cell (less females died than expected under the hypothesis of no association between GENDER and SURVIVAL). The opposite pattern holds true as well, as indicated by the large significant positive residuals.

Understanding the details of mosaic plots not only enhances our ability to interpret datasets visually but also provides a comprehensive view of relationships within the data. While the mosaic plot is particularly beneficial for visualising larger tables, the choice between using this plot and the table of standardised residuals is ultimately up to the analyst. It depends on their intention and the specific information they aim to convey.

3.2 Statistical Significance and Sample Size: Things to Consider

When we carry out our chi-squared test, we have to be aware of an important aspect that revolves around the significance of the result from a statistical point of view and its relationship with the sample size (cross-tab's grand total). As you should know by now, the *p*-value associated with the chi-squared statistic gives us clues as to how likely is that the two variables are independent in the population from which we drew our sample.

However, let's picture this. Imagine multiplying every cell of a cross-tab by a constant factor; it makes intuitive sense that we end up increasing the sample size proportionally. Since the chi-squared test is sensitive to sample size, as the sample size increases, the test becomes comparatively more powerful at detecting differences between the observed and expected frequencies. This is akin to using a magnifying glass: a larger sample size, like a magnifying glass, gives us more *power* to see how things are, but there could be the risk of considering big (or significant) what is, in reality, quite tiny.

This analogy helps us understand that, with larger samples, we might over-estimate the significance of our findings.

Let's have a look at Table 3.1. The chi-squared statistics for table A is equal to 0.105, with an associated *p*-value of 0.746. The test proves not significant, indicating that in the parent population SURVIVAL and GENDER are not likely to be associated. So far so good, right? However, let's have a look at table B, where we have multiplied every cell by 100. It is obvious to say, but the proportion of males and females in each SURVIVAL category is the same as in the preceding example. The same holds exactly true for the proportion of died and survived in each GENDER category. However, now the chi-squared statistic is (not surprisingly) 100 times larger (10.506), with an associated *p*-value of 0.0012, which proves statistically significant.

What we have seen here is that, even though the proportions are the same in both tables, the chi-squared test is more sensitive to any differences in the larger table due to the increased sample size. In the smaller table, the differences between the observed and expected frequencies did not prove large enough to attain statistical significance. However, when we multiplied each cell by 100, the differences became more pronounced. Let's think about this: the expected frequency in the upper-left cell of the small table is 9.5, whereas in the larger table is 950, again a huge (100-fold) rise in numerical terms. This led the chi-squared test to conclude that there is a significant association between SURVIVAL and GENDER. In practical terms, the above means that, when dealing with large sample sizes, you might find a significant association as a mere result of the table's grand total, even if the actual dependence is tiny and of scant substantive importance.

What is pictured above highlights two important aspects to consider. The first is that larger samples may result in the rejection of the null hypothesis of independence when the actual relationship is trivial. Secondly, the chi-squared test is only a preliminary tool for analysing the association between categorical variables. We have to always consider coupling the associated *p*-value with measures that can provide us with an idea of the strength of

TABLE 3.1

(A) Cross-Tabulation of SURVIVAL Outcomes and GENDER for a Fictitious Group of 38 Titanic Passengers, and (B) Obtained Multiplying A by 100

A	GENDER			B	GENDER		
SURVIVAL	Male	Female	Total	SURVIVAL	Male	Female	Total
Died	9	10	19	Died	900	1000	1900
Survived	10	9	19	Survived	1000	900	1900
Total	19	19	38	Total	1900	1900	3800

Note: The association in A is not significant (chi-squared value: 0.105; *p*-value: 0.746), in B is significant (chi-squared value: 10.506; *p*-value: 0.0012).

the association. What they help us with is to quantify the magnitude of the relationship regardless of the sample size. I cover this in Chapter 4.

3.3 Small Numbers, Big Questions

3.3.1 The Chi-Squared Test and Small Expected Frequencies in 2 × 2 Cross-Tabs

When the expected frequencies are very small, the chi-squared test might not yield accurate results. This inaccuracy arises because the test's mathematical foundation presumes a large number of observations. Historically, statisticians have proposed several rules of thumb for using the chi-squared test. Early guidelines recommended that all expected values be greater than 10, while some authors suggested a threshold of 20. However, these figures are often seen as too limiting. Some scholars advocate a less stringent criterion, suggesting that the chi-squared test is reliable if no expected counts are smaller than 5. Others argue that all counts should be equal to or larger than 1.

Yet other guidelines recommend that a single expected count can be as low as 0.5, provided that the remaining values are larger than 5 or 10. To complicate matters, some scholars suggest that the chi-squared test can be reliable if 80% of the expected counts exceed 5, and the remaining 20% fall between 1 and 5. Additionally, some posit that, as long as the total sample size exceeds 20, the number of categories is five or more, and all expected values are larger than 3, the chi-squared test remains reasonably accurate.

Now, if you have managed to keep up with all that without your head spinning, hats off to you! Quite fortunately, amidst that body of recommended thresholds, another scholarly perspective emerges, simplifying the application of the chi-squared test in situations where small expected frequencies are present. It proposes that the test's validity is not as fragile as once thought, especially when considering the average expected frequency across all cells in the cross-tab, rather than the minimum expected value in any single cell.

The guideline suggests that an average expected frequency of at least 5 or 6 across all cells in the table is sufficient for maintaining the chi-squared test's reliability at the 0.05 significance level. So, how will we know beforehand whether our cross-tab features an average expected frequency of at least 5 or 6? The answer is pretty simple: the table's grand total is divided by the number of cells (that is, the number of rows times the number of columns). Just to provide you with an actual example, in our real Titanic data (Table 2.4), the average expected frequency is 1309 : (2 × 2) = 327. Should you want to verify the latter figure yourself, it suffices to do the simple math using the expected counts reported in the mentioned table: $(521 + 288 + 322 + 178) : 4 = 1309 : 4 = 327$. This happens for the simple reason that the sum of the expected counts is, quite logically, equal to the cross-tab's grand total.

The above has an interesting implication. It allows us to determine the minimum grand total necessary for any cross-tab to reach an average expected count of 5 or 6 and, therefore, to maintain the traditional chi-squared test's reliability. One has to just multiply 5 or 6 by the total number of cells. For a 2 × 2 table, this calculation would result in a minimum grand total of 20 or 24. Similarly, for a 2 × 4 table, the calculation yields a grand total of 40 or 48. So, if you analyse a cross-tab of size 2 × 2 (featuring a grand total of *at least* 20 or 24) or a cross-tab of size 2 × 4 (featuring a grand total of *at least* 40 or 48), the chi-squared test is bound to yield reliable results.

While this approach offers a criterion that cuts through the confusion of multiple rules, providing a coherent threshold, it is still prudent to consider other strategies for situations where these conditions are not met. What if, for example, we need to analyse a 2 × 2 cross-tab with a grand total smaller than 20 (or 24), or a 2 × 4 table with a grand total smaller than 40 (or 48)? Indeed, there are viable options for these scenarios, which we explore in the following sections. A summary of the different approaches is provided in Section 3.5, along with tentative schematic guidelines as to the choice of a suitable analytical approach given the cross-tab's size, grand total, and expected counts.

3.3.2 The $(N - 1)/N$ Correction

For 2 × 2 tables with small expected frequencies, a simple correction adjusts the chi-squared statistic by multiplying it by $(N - 1)/N$, where N represents the total sample size. This adjustment is especially pertinent for smaller samples (N less than 20) with all expected counts equal to or larger than 1. While scholars have studied this correction in great detail (have a look at the readings suggested in Section 7.4.4), your humble author could not resist diving in and running the simulations himself. Why, you ask? Well, some people climb mountains because they are there; I run statistical simulations because ... I can. Plus, a bit of recreational number-crunching never hurt anyone. Call it a peculiar passion!

The chi-squared test, as with any statistical test, has the inherent risk of leading us to the wrong conclusion, indicating that a dependence exists when it actually does not. By convention, as we saw in Section 2.3.4, this risk is set at a probability of 0.05 (5%). But with small samples, this risk can increase. The mentioned correction factor, $(N - 1)/N$, should reasonably keep that risk close to the predetermined level (typically, 5%), especially when dealing with small samples. My simulation helped me have a first-hand experience with the effectiveness of this correction.

What did I do? I conducted 10,000 simulations for sample sizes ranging from 5 to 50, at 1-unit increments, though it is worth noting that the lower end of this range is certainly not a realistic sample size for most research applications. For every iteration, the software I used (1) generated a random 2×2 table featuring no true association (independence); (2) calculated the chi-squared test statistic for this table, both in its traditional form and with the $(N - 1)/N$ correction; (3) noted whether each test flagged a statistically significant association (at the typical 5% significance level). The key outcome to take note of was the significance rate, that is, the proportion of simulations (out of the 10,000 for each sample size) where each test indicated a significant result. A rate around 5% would be expected by chance alone, given that the simulated datasets were independent. Thus, rates significantly higher or lower than 5% would indicate an overestimation or underestimation of the significance by the tests, respectively. This provides insights into the tests' reliability across different sample sizes. In total, this process resulted in a total of 460,000 simulated tables, not to mention a number of coffees consumed while anxiously waiting for the simulation to complete.

What results did I get? Let's have a look at Figure 3.4. At the smallest sample size of 5, where both the minimum and average expected counts were considerably low (minimum 1.89, average 1.25), the traditional test proved *liberal*, indicating an association about 6.1% of the time, exceeding the desired

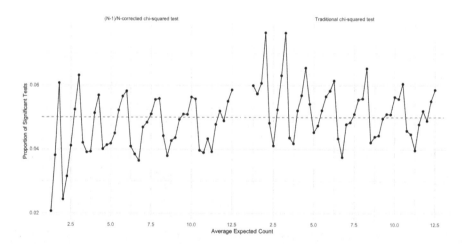

FIGURE 3.4
Line chart illustrating the differing rates at which the traditional chi-squared test (right) and its $(N - 1)/N$ corrected version (left) identify significant associations across simulated 2×2 tables of varying grand totals (N ranges from 5 to 50, inclusive) and average expected counts (1.25–12.5). The average expected count is equal to N divided by 4, which is the number of cells (see Section 3.3.1). Note the traditional test's liberal results at smaller grand totals, corresponding to lower average expected counts, compared to the more conservative nature of the corrected test. As the average expected counts increase, indicating larger sample sizes, a convergence in the performance of both tests is observed. The chart is based on the results of a simulation, with details provided in the text.

5% level. In contrast, the $(N - 1)/N$ corrected version was more *conservative*, suggesting an association only about 2.1% of the time. As the sample size increased to 10 (minimum expected count 4.16, average 2.5), a convergence in the performance of both tests was observed. They indicated associations at a similar rate of approximately 4.1%, aligning more closely with the 5% target.

For sample sizes ranging from 11 to 20 (corresponding to a minimum expected counts between 4.58 and 8.74, and an average expected counts between 2.75 and 5.00), the difference in the performance of the corrected and traditional chi-square tests continued to diminish. This trend was evident as both tests continued to show more similar rates of significant results. Particularly at a sample size of 20, representing the higher end of this range, this convergence became more pronounced. At this point, with a minimum expected count of 8.74 and an average expected count of 5.00, the rate of significant results for the corrected and traditional test was observed to be 4.18% and 4.53%, respectively. This specific observation at a sample size of 20 highlights that, by the upper limit of this sample size range, the distinction between the two tests in terms of their rates of detecting significant results is minimal.

As N further increased above 20, with average expected counts larger than 5, the term $(N - 1)/N$ in the corrected test approached 1, making the adjusted chi-squared value nearly equivalent to the uncorrected value. For instance, at a sample size of 50, where the minimum expected count was 22.99 and the average expected count was 12.5, both tests showed a significance rate of around 5% (5.86% to be precise), demonstrating their alignment at higher sample sizes.

What did we learn? The $(N - 1)/N$ corrected chi-squared test generally gives results that are comparatively more in line with what we would expect by chance alone (about 5% of the time) for most sample sizes. This suggests that this test is balanced, neither being too quick to suggest that something significant is happening when it is not (not too liberal), nor being too cautious and missing something important (not too conservative). On the other hand, the traditional chi-squared test (in cases with small samples, low expected counts, and average expected counts under 5), seems more prone to indicating significant results. This means that this test might be a bit too quick to suggest that something interesting is happening when it might just be a random occurrence.

All in all, it seems beneficial to use the corrected version of the chi-squared test for cross-tabs with a grand total smaller than 20 and expected counts equal to or larger than 1. For large samples, the results of the two tests converge because, as said, as N increases, the term $(N - 1)/N$ approaches 1. In fact, in our Titanic example, the traditional chi-squared value was 365.89, whereas the corrected one is $365.89 \times (1309 - 1) : 1309 = 365.89 \times 0.99 = 365.61$.

Armed with these insights, let's turn our attention to an example to see the correction in action. Consider the hypothetical dataset in Table 3.2, which has a sample size of 17, and an average expected count of $17 : 4 = 4.25$.

Specifically, two expected counts approximate 5 (Male-Died: 5.82; Male-Survived: 5.18), and two others lie between 1 and 5 (Female-Died: 3.17; Female-Survived: 2.82). The traditional chi-squared statistic is 4.89, with an associated p-value of 0.0269. The test indicates a significant dependence. After applying the correction, the chi-squared value adjusts to $4.89 \times (17 - 1): 17 = 4.89 \times 0.94 = 4.60$, and its p-value rises to 0.032. The result is slightly more conservative, but still points to a significant association (see further comments in Section 3.3.5).

3.3.3 The Fisher's Test

The Fisher's test is a common alternative to the chi-squared test when small expected frequencies are encountered. While the chi-squared test measures the difference between observed and expected frequencies, Fisher's test (which, by the way, is named after the eminent British statistician R. A. Fisher) adopts a unique perspective.

Let's consider our previous dataset featuring small expected counts. Given the fixed marginal frequencies of the cross-tab (I will come back on this in the next section; bear with me), the test calculates the probability of observing our specific table under the assumption of no association between GENDER and SURVIVAL. It does this by considering all possible tables that could arise with these margins.

Think of each possible cross-tab as a unique outcome. Some of these outcomes are more likely than others under the null hypothesis of independence. While the chi-squared test compares the chi-squared statistic (which distils down into a number the discrepancy between the observed and the expected counts) to the distribution of that statistic under the null hypothesis to see how extreme the observed chi-squared value is (we covered this in Chapter 2), the Fisher's test works differently. It compares our observed outcome (the cross-tab under analysis) to all other possible tables that can be generated keeping fixed the marginal sums, to see how extreme it is. For

TABLE 3.2

Cross-Tabulation of SURVIVAL Outcomes and GENDER for a Fictitious Small Sample of 17 Titanic Passengers

SURVIVAL	GENDER		
	Male	Female	Total
Died	8	1	9
Expected	5.82	3.17	
Survived	3	5	8
Expected	5.18	2.82	
Total	11	6	17

each possible table, the test calculates the probability of observing that table using what is called *hypergeometric distribution*. The p-value of our table is determined by summing the probabilities of all tables that are as unlikely or more unlikely than our observed table.

Now, you might be wondering about the hypergeometric distribution. In layman's terms, imagine you have a big bowl of mixed fruit, like apples and oranges. If you were to randomly pick a certain number of fruits without putting them back, the hypergeometric distribution would tell you the likelihood of getting a specific number of apples and oranges. In our Titanic example, it is like asking: 'Given the total number of males, females, survivors, and non-survivors, what is the chance of observing the specific counts in our table?'.

The formula for this probability is

$$p = \frac{(a+b)!(c+d)!(a+c)!(b+d)!}{N!a!b!c!d!}$$

where a, b, c, and d represent the cells of a 2x2 cross-tab (a being the top-left one, d the bottom-right one), and N represents (as you know) the table's grand total (or sample size). By the way, the exclamation point is a mathematical notation (called *factorial*) that indicates multiplying a given number sequentially by each positive integer less than that number. For instance, 3! is equal to $3 \times 2 \times 1 = 6$. Fisher's test gives the probability of observing our specific table out of all the tables that can be formed given the marginal totals.

Plugging in the values from our Table 3.2, we have

$$p = \frac{(8+1)!(3+5)!(8+3)!(1+5)!}{17!8!1!3!5!} = \frac{9}{221} = 0.041$$

The test proves significant, with a p-value that is larger than the one associated with the traditional chi-squared test (0.0269). The Fisher's test p-value proves a bit larger than the one associated with the corrected chi-squared statistic (0.032). Both are more *conservative* compared to the result of the traditional chi-squared test. I comment further on this in Section 3.3.5.

Before diving into the next discussion, you might be wondering how Fisher's test differs from the Monte Carlo simulated p-value method that we explored in Section 2.3.5. Both indeed deal with tables generated under the null hypothesis and even keep the marginal totals fixed. The key distinction lies in the approach: Fisher's test exhaustively computes the exact probabilities for *all* possible tables given the fixed margins. In contrast, the Monte Carlo method produces a large, but *finite*, number of newly generated random tables to create the distribution of the test statistic under the null hypothesis. It then compares the observed chi-squared statistic to this generated distribution to yield a p-value. While the Monte Carlo method may prove useful when analysing larger tables (see Section 3.4.4), it could also represent another viable option for small tables as well, together with another approach (namely, the permutation method) that I cover in Section 3.4.4.

3.3.4 Locked Margins, Unlocked Secrets: The Fisher's Test Debate

In elaborating on Fisher's test, we saw that there is an underlying assumption: the marginal totals are considered fixed, meaning that the total number of occurrences in each level of the two variables is predetermined. This implies that the range of possible tables that the test considers, while calculating probabilities, all adhere to these specific marginal totals. However, this might not always align with the experimental design or data collection methodology, as a number of statisticians pointed out.

Imagine a research study examining the relationship between smoking status and lung cancer across various geographical regions. Researchers gather data from different hospitals and clinics, totalling the instances of lung cancer in smokers and non-smokers. In this instance, the marginal totals (the total number of smokers and non-smokers, and the total number of cases with and without lung cancer) may not be fixed because they were not predetermined before the study but observed from collected data. The study's data comes from natural observation rather than a structured experimental design that enforces certain totals. In a scenario where the marginal totals are not predetermined, using a test that assumes fixed marginals, like Fisher's test, might not be entirely apt.

Though it might seem like a straightforward discrepancy to resolve with the power of modern computational facilities, navigating through the literature on fixed and random marginals swiftly unveils the depth and intricacy of the debate. However, this is but one facet of the various debates surrounding Fisher's test, and a comprehensive review is beyond the scope of this book. If you find yourself in a situation where the assumption of fixed marginals might not hold, or if the general subject sparks your interest, the works cited in Section 7.4.4 serve as a starting point in grappling with these statistical nuances.

3.3.5 $(N - 1)/N$ Correction and Fisher's Test: Things to Consider

On the basis of the data in Table 3.2, we saw that both the $(N - 1)/N$ correction and the Fisher's test indicated a significant association, and the same was indeed doing the traditional chi-squared test (but with a comparatively smaller p-value). You might wonder: 'So, what's the big deal in using alternative tests?'.

Imagine you are a doctor deciding on a treatment approach. The traditional, uncorrected, chi-squared test might confidently point you towards a specific treatment. But when you apply the $(N - 1)/N$ correction or the Fisher's test, the evidence might not seem as robust. Sometimes, their p-value might even suggest the treatment is not as effective as initially thought. Why fuss over these nuances? Because these *p*-values often drive decisions, not just in medicine but across various fields. Whether it is launching a new medical treatment, reshaping an educational policy, or even making judicial decisions, these values play a pivotal role. And a more conservative *p*-value, like the ones from the corrected chi-squared or Fisher's test, might be the deciding factor. It is a reminder to tread carefully, especially with smaller samples.

Reflecting on what we have seen, and mindful of the debates on the Fisher's test, both the latter and the $(N - 1)/N$ corrected chi-squared test emerge as more cautious alternatives to the traditional chi-squared test, especially with small samples. Their comparatively larger *p*-values are a testament to this caution. And while some statisticians recognise that Fisher's test might be a bit too conservative (being too cautious and accepting the null hypothesis too often), it is essential to remember the context. With small samples and (more importantly) with small expected counts, the traditional test can sometimes jump the gun, seeing things that might not be there. In contrast, the two reviewed alternatives offer a more cautious view, ensuring we are not just seeing patterns because of small numbers but genuine trends.

For researchers and decision-makers, the message is clear: when data is limited, it is wise to lean on more cautious tests.

3.4 Addressing Small Expected Frequencies in Larger Tables

3.4.1 Introduction

Even though I elaborate on analytical approaches to larger tables later on in this book (Chapter 7), to maintain a coherent and tight focus on the problem of the small expected frequencies, this section elaborates on the issue in relation to cross-tabs larger than 2 × 2. When dealing with such tables, the strategies to address small expected counts are somewhat diverse. Some researchers stand by the traditional chi-squared test, advocating its reasonable accuracy provided that the sample size is over 20 or (as seen) that the average expected count is larger than 5 or 6. These studies have shown that the traditional test can be surprisingly robust, even in scenarios with small cell expectations in larger tables. However, other approaches are indeed available.

In particular, we are going to familiarise ourselves with the new possibilities that the advancement of computational methods has opened up (Section 3.4.4). These approaches, specifically permutation and Monte Carlo methods (we have met the latter in Section 2.3.5), leverage the power of randomisation and computational simulations to provide alternative ways to test the significance of the association in our observed data. These methods do not modify the existing data or apply a correction to the chi-squared statistic; they take a different path by creating a multitude of potential scenarios through randomisation. These approaches allow us to assess the robustness of our observed association against a backdrop of what could happen purely by chance. It is akin to testing our data against a universe of possibilities, ensuring that our conclusions are not just a fluke of sample sizes or an artefact of our test conditions. By examining these methods, we will gain a deeper appreciation for their ability to strengthen our confidence in the statistical associations we observe, especially when traditional methods may fall short.

3.4.2 Pooling Levels

For cross-tabs larger than 2 × 2, a traditional strategy you can use (even though, admittedly, it might prove not always viable and/or advisable; see later on) is to combine levels to create larger expected frequencies. For example, if you are looking at people's favourite fruit and you have levels for "apples", "bananas", "oranges", and "other", you could combine "apples" and "bananas" into one category, and compare the combined "apples and bananas" category to "oranges" and "other". This would increase your expected frequencies and make the chi-squared test more reliable.

However, some statisticians warn against pooling levels for a number of reasons. By merging levels, we can lose details about our data with the consequence of losing potentially useful information. The procedure may be also thought of as violating the randomness of the sample we deal with. The chi-squared test, as well as any other statistical test, hinges on the assumption that our sample is a random slice of a larger population and that, as a consequence, there are certain probabilities for our observations to fall into the levels of the two variables under study. Pooling levels after the data have been drawn may alter those probabilities, causing our sample to deviate from its parent population.

3.4.3 (N − 1)/N Correction and Fisher's Test for Larger Cross-Tabs

What about applying the $(N - 1)/N$ correction? Admittedly, I did not find conclusive evidence on this from the literature. My personal take is that in high-stakes contexts (for instance, medical trials, legal decisions, or policy changes), the cost of falsely rejecting a true null hypothesis can be extremely high. In such cases, being conservative and minimising such risk may be prioritised, even if the correction has (as we saw) a diminishing effect as N grows. Applying the correction, even when N is large, may prove a way to exercise this conservatism systematically. Further to that, there may be situations where the p-value is very close to the significance level. In these cases, even a small adjustment from the $(N - 1)/N$ correction could tip the balance between rejecting and not rejecting the null hypothesis. In high-stakes situations, this adjustment (however small) could be seen as a safeguard against potential false positives.

What about the Fisher's test for larger tables? Fisher's test is primarily designed for 2 × 2 tables. It can be computationally intensive for larger tables, given the need to enumerate all the possible cross-tabs as extreme as the observed one. This computational intensity scales disproportionately as the table size increases, making it less feasible for larger tables. When computational resources are available, it might be worth considering.

While the computational demands of Fisher's test increase with larger tables, necessitating significant resources, the following methods, though also reliant on computational power, offer a scalable and potentially more feasible approach for assessing statistical significance in such contexts.

3.4.4 Permutation and Monte Carlo Methods

The permutation approach emerges as a suitable alternative method for large tables. This approach extends the testing principles seen in Fisher's test, but instead of evaluating all possible table configurations, it involves randomly shuffling a given number of times the observed data within the cells of the cross-tab, keeping the marginal totals constant. That does not sound clear, does it? Let me illustrate the underlying idea with an example. Even though I am indeed elaborating on the use of the method for large tables, to keep things simple, a 3 × 3 cross-tab is used.

Have a look at Table 3.3. The table contains a fictitious list of 20 Titanic passengers for whom we have information about CLASS and EMBARKATION PORT. We already encountered this type of data organisation earlier on, in Table 1.1. For the time being, disregard the last column to the right. We might be wondering whether an association is likely to exist between the two variables. For example, we might be seeking to understand if there was a sort of social-class bias between people living in those three cities when it comes to the fare they could afford when buying the ticket. Therefore, we want to test for a dependence between CLASS and EMBARKATION PORT.

When we cross-tabulate the data in Table 3.3, we get the cross-tab represented in Table 3.4A. The table features small expected counts: they are all below 3, and the ones for the second class are 1.75, 1.50, and 1.75, respectively.

The logic underlying the permutation-based method is the following: under the null hypothesis of no association, we assume that the CLASS a passenger travelled in is independent of their EMBARKATION PORT. Consequently, each level of the EMBARKATION PORT variable becomes merely a label that can be randomly reassigned (or permuted) without changing the inherent probabilities of our observed data. Therefore, we reshuffle (or permute) the EMBARKATION PORT labels, effectively breaking any existing link between CLASS and EMBARKATION PORT to reflect the null hypothesis of no association. This single random permutation leads to a new dataset, which is represented by the second and fourth columns in Table 3.3. This

TABLE 3.3

Summarising the CLASS of a Fictitious Group of 20 Titanic Passengers, Categorised by EMBARKATION PORT

Obs. No.	CLASS	EMBARKATION PORT	RANDOM PERMUTATION
1	2nd	Southampton	Southampton
2	1st	Cherbourg	Cherbourg
3	3rd	Queenstown	Queenstown
4	2nd	Southampton	Cherbourg
5	2nd	Cherbourg	Southampton
6	3rd	Queenstown	Southampton
7	3rd	Southampton	Queenstown
8	3rd	Queenstown	Southampton
9	3rd	Queenstown	Queenstown
10	1st	Southampton	Southampton
11	1st	Cherbourg	Queenstown
12	1st	Queenstown	Cherbourg
13	2nd	Southampton	Cherbourg
14	2nd	Cherbourg	Queenstown
15	1st	Queenstown	Cherbourg
16	3rd	Southampton	Queenstown
17	3rd	Southampton	Southampton
18	3rd	Queenstown	Southampton
19	1st	Cherbourg	Cherbourg
20	1st	Cherbourg	Queenstown

Note: The last column to the right contains a random permutation of the EMBARKATION PORT labels, illustrating a possible rearrangement under the null hypothesis that passenger CLASS is independent of the EMBARKATION PORT (see also Table 3.4B).

new dataset, which is as equally likely as the original under the null hypothesis, is then used to construct another cross-tab (Table 3.4B) on which the chi-squared statistic is computed. In this permuted dataset, the chi-squared value is 3.37.

By repeating this permutation process multiple times, we build a distribution of chi-squared statistics that we would observe under the null hypothesis of independence. The chi-squared statistic (9.61) from the original cross-tab (Table 3.4A) is then assessed against this permutation distribution. The proportion of permuted tables that yield a chi-squared value as extreme as or more extreme than our observed value gives us the permutation p-value. In our example, 426 out of 10,000 permutations yielded a chi-squared value equal to or larger than 9.61. Therefore, with a *p*-value of 0.0426, the permutation method indicates that there is a significant association between CLASS and EMBARKATION PORT. It is important to note,

TABLE 3.4

(A–B) Cross-Tabulations of CLASS and EMBARKATION PORT for a Fictitious Sample of 20 Titanic Passengers

A

CLASS	EMBARKATION PORT			
	Cherbourg	Queenstown	Southampton	Total
1st	4	2	1	7
2nd	2	0	3	5
3rd	0	5	3	8
Total	6	7	7	20

B

CLASS	EMBARKATION PORT			
	Cherbourg	Queenstown	Southampton	Total
1st	3	2	2	7
2nd	0	2	3	5
3rd	3	3	2	8
Total	6	7	7	20

Note: (A) Cross-tab based on the data from Table 3.3 (chi-squared value: 9.61). (B) Cross-tab derived from Table 3.3, but using the permuted embarkation port labels (chi-squared value: 3.37). Both cross-tabs feature the same marginal totals. The comparison serves to illustrate the differences that arise in the distribution of passengers across classes and ports due to random permutation of the EMBARKATION PORT labels in the original Table 3.3.

as we did in Section 2.3.5, that since the permutation process is random, repeating it n times can yield slightly different permutation distributions and, consequently, p-values. However, these variations are typically minor and the overall conclusion about the significance of the association should remain stable across iterations.

It is worth noting that we can use the Monte Carlo-based chi-squared test, which we touched upon in Section 2.3.5, in the same situations where the permutation method can be put to work. Both are generally more computationally manageable than the Fisher's test for larger tables. The Monte Carlo method, as you may remember, involves simulating a large number of tables based on observed marginal totals and calculating the chi-squared statistic on each. For the data in Table 3.4A, applying the Monte Carlo method resulted in 397 (out of 10,000) random cross-tabs yielding a chi-squared value equal to or larger than 9.61. This corresponds to a p-value of 0.0397, pointing to a significant association as well.

At this point, before ending this section, it is essential to distinguish the permutation from the Monte Carlo method because you are surely wondering what is the difference between the two. The difference is admittedly subtle and indeed needs clarification, especially for an audience of novices in the field. While the permutation test reshuffles the data the way we described, which is tantamount to reorganising the actual observed data within the cross-tab's cells (keeping the marginal totals unchanged), the Monte Carlo method generates completely new data while adhering to the observed marginal totals. Each approach culminates in a chi-squared statistic used to derive a p-value, yet they fundamentally differ in their data handling. The permutation test directly assesses the observed data's significance within the context of its original structure as opposed to the Monte Carlo method, which creates fresh datasets.

3.5 Choices in Chi-Squared Testings with Small Expected Frequencies

As we conclude this chapter, summarising what we have covered in the preceding Sections 3.3 and 3.4 regarding the issue of small expected counts in chi-squared testing can prove useful. When dealing with small expected frequencies, we have seen that a variety of alternatives are indeed available.

For 2×2 tables, both Fisher's test and the $(N - 1)/N$ correction offer valuable solutions (Sections 3.3.2–3.3.5). Fisher's test provides accurate results (though it is considered overly conservative by scholars) whereas the traditional chi-squared test may not perform well due to low expected frequencies (smaller than 1). The $(N - 1)/N$ correction serves as a subtle and effective adjustment in situations where the cross-tab grand total is smaller than 20 and the expected frequencies are equal to or larger than 1. It can also be used on tables of any size as well.

We have to consider that, before giving up on the traditional chi-squared test, we can preliminarily apply the average expected frequency rule (Section 3.3.1) to ascertain whether the average expected count across the entire cross-tab is sufficient for maintaining the chi-squared test's reliability at the 0.05 significance level. This rule helps to gauge the overall adequacy of the expected counts in providing reliable results from the traditional chi-squared test.

The same check would be reasonable not only for 2×2 cross-tabs, but for larger tables as well. For these, the permutation and Monte Carlo methods (described in Section 3.4.4) represent two suitable options, offering robust approaches through computational simulations. These methods are particularly beneficial when the expected counts are too small for a reliable traditional chi-squared test application (or for carrying out the $(N - 1)/N$ version of the test), but the cross-tab under analysis proves too large for the practical use of Fisher's test.

Needless to say, given the widespread availability of computational resources today, simulation-based methods such as permutation and Monte Carlo tests are easily accessible and can be applied to a broad range of situations, including those where the traditional chi-squared test is also a viable option. Nonetheless, the use of these methods should be considered in the context of the researcher's familiarity with such approaches and the computational demands of the specific analysis.

To provide a structured (and admittedly tentative) approach to the selection of an appropriate strategy, Figure 3.5 presents a decision flowchart. The chart is intended as a tentative guide. The sequential steps, while beneficial in their clarity, are far from matching the depth and flexibility of the thoughtful decision-making one must engage in when dealing with chi-squared testing. The flowchart should be recognised as a simplification for educational and practical purposes. While it provides a structured approach, the decisions it leads to are based on guidelines and thresholds that may not apply to every dataset or research question. The chart serves as a sort of frame of reference, not as an absolute rule. The analyst's judgment is required to determine the best course of action, especially in borderline cases.

In conclusion, the body of alternatives that we have reviewed enables us to tackle the issue of small expected counts in chi-squared testing with some confidence, allowing to select a suitable method based on the size and other features of our cross-tabs. However, we have to bear in mind that each method comes with its own set of assumptions and nuances. The key is to approach the chi-squared test and its alternatives with a critical eye, understanding the underlying logic as well as the potential limitations, and making an informed choice that is methodologically sound and appropriate given the context of the analysis.

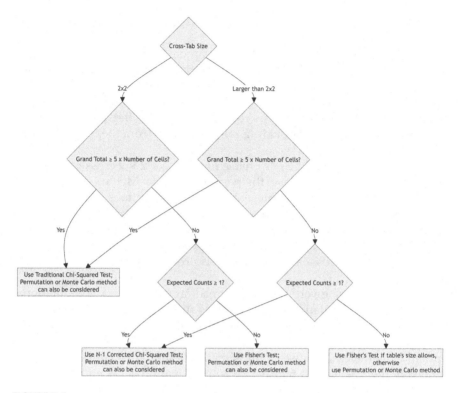

FIGURE 3.5
Tentative decision flowchart for selecting an appropriate approach to chi-squared testing when small expected frequencies are present. This diagram tentatively guides through a series of choices based on the size of the cross-tabs, the grand total, the expected counts, and the average expected count. While the flowchart aims at providing clarity and direction, it cannot replace the analyst's judgment required to determine the best course of action.

3.6 Key Takeaways

- Adjusted standardised residuals help identify which specific cells in a contingency table are significantly contributing to the chi-squared statistic.

- Cells with absolute values of adjusted standardised residuals larger than 1.96 are those that differ significantly from what would be expected under independence.

- Mosaic plots offer an intuitive way to visualise chi-squared residuals, revealing patterns and relationships.

- Tile attributes like height, width, and colour in a mosaic plot provide insights into proportions, overall sample distribution, and statistical significance, respectively.

- The chi-squared test is sensitive to sample size. Even if the proportion of categories remains consistent, larger sample sizes can cause minor differences to appear statistically significant.

- It is crucial to consider both the p-value and measures of the strength of association between categorical variables to understand the depth and relevance of the relationship.

- The chi-squared test's reliability can be compromised when expected frequencies are too low.

- There exist various guidelines for minimum expected frequencies, such as all expected values should be above 5. However, these guidelines are not universally accepted, highlighting the importance of exercising judgment.

- The average expected frequency rule suggests that an average count of 5 or 6 across all cells of a cross-tab is generally sufficient to maintain the reliability of the chi-squared test at a 0.05 significance level, offering a more flexible approach in the face of small expected frequencies.

- Merging categories can be a strategy to increase expected frequencies and make the chi-squared test more reliable, though it may not always be feasible/advisable.

- For 2×2 tables with a sample size smaller than 20, corrections like the $(N - 1)/N$ adjustment or the Fisher's test can be used to reduce the risk of rejecting the null hypothesis when it is actually true.

- While the chi-squared test measures the difference between observed and expected frequencies, Fisher's test calculates the probability of observing a specific table, contrasting it with all potential tables.

- In larger tables, the application of the $(N - 1)/N$ correction can still be a prudent approach, especially in high-stakes contexts. This in spite of the fact that as N grows, the corrected and uncorrected chi-squared statistics tend to converge.

- Fisher's test is primarily designed for 2×2 tables. While the test can be extended to larger tables, it can be computationally intensive and might not always be the most practical choice for cross-tabs larger than 2×2.

- The Monte Carlo-based chi-squared test generally offers a more computationally manageable alternative. It can be particularly advantageous for larger tables, as it typically scales better with size and complexity, making it a more practical choice.

- The permutation approach offers another computationally feasible alternative to Fisher's test for larger tables.

- Permutation tests maintain the data structure and shuffle labels within the table to generate a distribution of the chi-squared statistic, providing an empirical p-value that reflects the significance of the observed association. This approach aligns with the testing philosophy of Fisher's test while overcoming its computational limitations.

- When comparing the permutation method to the Monte Carlo-based chi-squared test, it is important to recognise that the former reshuffles actual observed data, whereas the latter generates new datasets based on observed marginal totals.

- Especially when working with small frequencies, it is vital to justify the chosen approach and be aware of potential pitfalls.

4

Strength in Numbers: Measuring Dependence

4.1 Chi-Square-Based Measures of Association

As seen in Chapter 3, the chi-squared statistic in its own does not tell us anything special, of course unless we use it to work out the associated p-value. But that is another (important) story. The chi-squared statistic cannot be used to measure how strong (or weak, or moderate) is the dependence between the two categorical variables being analysed.

What do I mean by that? Let's consider two things: (1) the existence of a dependence between GENDER and SURVIVAL (to resume our example) does not necessarily mean that the relation is strong: one thing is knowing that you like ice cream, another thing is knowing how strongly you crave for it; (2) even though the chi-squared statistic summarises the amount of divergence between the observed and the expected counts, its magnitude is influenced (as seen in Section 3.2) by the size of our cross-tab and by how large are the counts involved. Other things being equal, larger tables will produce large chi-squared statistics. By the same token, two equal-sized tables featuring counts of different magnitudes will produce different chi-squared values. This is why the chi-squared statistic on its own cannot be employed as a measure of the strength of the dependence (if any). It would be like trying to give drivers a fine for overspending, but the faster a car is, the larger the speed limit rises. We would never be able to give a fine to any driver.

There are a number of measures of association that we can use. In the following sections, I will introduce you to well-known ones.

4.1.1 Contingency Coefficient

The first measure of association one can use is the contingency coefficient, which is typically indicated with C. Based on the chi-squared statistic, it is calculated as follows:

$$C = \sqrt{\frac{\chi^2}{\chi^2 + N}}$$

DOI: 10.1201/9781032726328-4

Shall I break that down for you? The chi-squared value is divided by itself plus the table's grand total (N), and the square root of the result is taken. What range of values can C assume? The minimum value it can attain is 0.0, when there is independence between the two variables being analysed. Since N cannot be (quite logically) 0, C will always be less than 1.0 or, to put it the other way around, C can never be equal to 1.0. This happens even when all the counts fall in one of the diagonals of the cross-tab. As an example, have a peek further down at Table 4.3A, where this configuration of counts is called *absolute* association and where C is 0.707.

To correct this deficiency, C is adjusted by dividing it by the maximum it can theoretically achieve given the table's size. The maximum can be worked out by the following formula:

$$C_{max} = \sqrt{\frac{k-1}{k}}$$

where k is the number of rows or columns, whichever is smaller. Once we know C_{max}, the adjusted version of C can be easily arrived at as follows:

$$C_{adj} = \frac{C}{C_{max}}$$

Note that if we want to compare the strength of association between tables of different sizes, only C_{adj} allows for a meaningful comparison since the adjustment takes care of the difference in sizes.

In the case of the Titanic dataset, C is equal to $\sqrt{\dfrac{365.89}{365.89 + 1309}} = 0.467$, C_{max} is equal to $\sqrt{\dfrac{2-1}{2}} = 0.707$, and C_{adj} is 0.467 : 0.707 = 0.660.

How shall we interpret the coefficient? It is not easy to attach labels to values of C (as well as to values of any coefficient in general, including the ones detailed in the following sections) because what can be considered weak (small), moderate (medium), or strong (large) is indeed field-specific. To touch upon a topic that we cover later on in Section 4.1.6, scholar J.F. Healey has proposed the following thresholds and related labels: *weak* dependence 0.0–0.10; *moderate* dependence 0.11–0.30; and *strong* dependence 0.31–1.0. The value of C_{adj} for our Titanic data points to an association that can be labelled as *strong*. However, it is important to note that these labels and thresholds are subject to debate. For other thresholds, have a look at the mentioned Section 4.1.6. For a brief reflection on the limitations and criticisms of chi-squared-based measures of association, you may refer to Section 4.1.7.

4.1.2 Phi Coefficient

The second measure of association that I would like you to familiarise with is the ϕ *coefficient* (pronounced *phee*; it is a Greek letter). To address the mentioned limitations of the chi-squared statistic as a measure of the strength of the dependence, ϕ divides the chi-squared value by the cross-tab's grand total and then it extracts the square root of the resulting quotient:

$$\phi = \sqrt{\frac{\chi^2}{N}}$$

The rationale for dividing the chi-squared statistic by the table grand total (N) is to normalise the measure and make it independent of the sample size. In fact, the chi-squared statistic, as noted in Section 3.2, increases with sample size, even if the association between the two variables remains the same. Why taking the square root? Recall that, for the chi-squared statistic to be calculated, the process required squaring the differences between the observed and the expected counts (see Section 2.2). Well, taking the square root effectively reverses that previous squaring, bringing the measure back to a scale that is akin to the original data scale. By dividing the chi-squared statistic by the table grand total and by extracting the square root of the result, we obtain a measure that reflects the strength of the association between the two variables regardless of the sample size. This means that ϕ remains the same if the total sample (that is, the cross-table's grand total) gets larger, as long as all the cells' count change by the same amount relative to each other. In the case of our real Titanic dataset (Table 2.3), the coefficient is 0.529.

The interpretation of ϕ is pretty much easy to grasp. In 2 × 2 cross-tabs, the coefficient spans from a minimum of 0.0 to a maximum of 1.0, which is attained only when all the counts fall in one of the diagonals of the cross-tab (have a peek again at the absolute association portrayed in Table 4.3A). A value near 0.0 means that the dependence is the weakest we can get, being practically non-existent. A value equal to 1.0 indicates that the dependence is the strongest possible.

Note that 1.0 is the maximum achievable value only when the table's marginal frequencies meet the following requirements: the sum of the first row is equal to the sum of the first column, or the sum of the second column is equal to the sum of the second row. When these conditions are not met, the coefficient's upper boundary is less than 1.0. To address this limitation (here we go again!), a *corrected* version has been proposed. This adjusted metric is described in the following section. Another modification (referred to as V) was introduced by a Swedish statistician in order to extend the use of ϕ to cross-tabs larger than 2 × 2. This is elaborated on in Section 4.1.4.

Before moving to the next section, I wanted to give you a heads-up. You might sometimes come across ϕ^2 in the literature. This value is the squared ϕ coefficient, that is, the value you obtain by simply dividing the chi-squared statistic by the table's grand total, without taking the square root. Do not let this different expression perplex you; in this scenario, you are encountering a distinct concept. The ϕ^2 provides us with the proportion of *variance*, the amount of variability in the observed counts that can be attributed to the association between the variables. In other words, ϕ^2 expresses the proportion of the variability in the observed frequencies that can be explained (or accounted for) by the relationship between the variables under analysis.

The concept of variance is typically associated with continuous variables, and might admittedly appear obscure in the context of cross-tabs. Think of the variance as the spread of observed frequencies around the frequencies we would expect under the hypothesis of no association. So, ϕ^2 can be understood as offering a sort of standardised insight into the departure (difference) between the observed counts and what we would expect if there were no association between the variables. This is the job we left in the hands of the chi-squared statistics, remember? At the end of the day, ϕ^2 is nothing but the chi-squared statistics divided by the table's grand total, but (unlike the chi-squared statistic) it is measuring the said departure in a standardised fashion, that is, independent of sample size. This measuring is done retaining the chi-squared metric that, as previously said, is not akin to our data scale.

Taking the square root is akin to bringing ϕ^2 back to our real world, expressing the standardised departure on a more intuitive metric, which can be interpreted in terms of strength, not variance. In our Titanic dataset, ϕ^2 would be the square of 0.529, which equals approximately 0.280. This suggests that about 28% of the variance in the observed frequencies of our dataset can be accounted for by the association between GENDER and SURVIVAL.

4.1.3 Limitations of Phi: Phi Corrected

To further refine ϕ, a corrected version known as ϕ_{corr} has been introduced. It accounts for the fact that the original ϕ coefficient may not have a maximum achievable value of 1.0 (as seen in the previous section) and, therefore, it is not directly comparable across tables featuring different marginals that do not meet the requirements mentioned earlier on. In such circumstances, in fact, different ϕ values are set within unique ranges that come with their own upper ceilings. To calculate ϕ_{corr}, one first computes ϕ_{max}, which represents the maximum possible value of ϕ under the given marginal totals. It is calculated using the following pretty scary formula:

$$\phi_{max} = \min\left(\sqrt{\frac{(a+b)\times(b+d)}{(a+c)\times(c+d)}}, \sqrt{\frac{(a+c)\times(c+d)}{(a+b)\times(b+d)}} \right)$$

where a, b, c, and d represent (as seen before) the cells of a 2 x 2 cross-tab (a being the top-left one, d the bottom-right one). If we stop scratching our heads and take a deep breath, it is easy to realise that the elements in the formula are essentially the marginal sums of the table. In other words, the individual row sums and column sums. For instance, $a + b$ is the sum of the first row, $a + c$ is the sum of the first column, and so on. As the formula tells us, ϕ_{max} is the smaller value that you get from those two calculations. The corrected ϕ is then computed as:

$$\phi_{corr} = \frac{\phi}{\phi_{max}}$$

This scaling relative to ϕ_{max} allows ϕ_{corr} to be a standardised measure, indicating how strong the observed association is relative to the strongest association that could possibly exist between those variables under the given marginal sums. In the case of our real Titanic dataset, ϕ is 0.529, as we previously saw. To come up with ϕ_{corr}, we first calculate ϕ_{max}:

$$\phi_{max} = \min\left(\sqrt{\frac{(682+127)\times(127+339)}{(682+161)\times(161+339)}}, \sqrt{\frac{(682+161)\times(161+339)}{(682+127)\times(127+339)}} \right)$$

If we do the math (again, take my word for it), ϕ_{max} is equal to 0.945. Our ϕ_{corr} is 0.529 : 0.945 = 0.559. As said earlier on, ϕ_{corr} measures the strength of the observed association relative to the strongest association that could possibly exist between the variables given the marginal sums. The strength of the relationship, compared to the maximum possible relationship (for the given marginal totals), comes up to about 56% in the Titanic case.

In our dataset, both ϕ and ϕ_{corr} can be considered as pointing to a *strong* dependence between GENDER and SURVIVAL. Also, note that in the case of the larger cross-tab previously taken into account (Table 3.1B) when elaborating on the relation between significance and sample size, the (uncorrected) ϕ coefficient is equal to 0.053. It points to a *weak* (almost negligible) association between the two variables, even though (for the reasons explained in Section 3.2) the chi-squared test proved significant. This highlights the importance of considering the strength of the association besides its statistical significance.

4.1.4 Cramér's *V* Coefficient

I need to warn you though. In cross-tabs larger than 2 × 2, ϕ can go past the upper ceiling of 1.0; it turns out to be difficult to interpret in that situation. For this reason, another chi-squared-based measure of association that we can use is Cramér's *V*. This coefficient, named after the Swedish statistician H. Cramér who devised it, is calculated as follows:

$$V = \sqrt{\frac{\chi^2}{N \times \min(r-1, c-1)}}$$

If you have a close look at it, you will notice that V is almost exactly similar to the (uncorrected) ϕ coefficient, with an extra term in the denominator. There, the cross-tab's grand total (N) is multiplied by the minimum between the number of rows (r) minus 1 and the number of columns (c) minus 1. Why?

Well, to answer this legitimate question I have to get into some technicalities, but it is really worth it.

Let's picture this: statisticians have shown that, for any given cross-tab, the maximum chi-squared value the cross-tab can theoretically attain is equal to $N \times \min(r - 1, c - 1)$. Therefore, the idea behind Cramér's V is to express the strength of the association as the ratio of the actual measured chi-squared value to the maximum it could achieve given the cross-tab's grand total and size. This makes sure that V is exactly bounded between 0.0 and 1.0. In fact, if the measured chi-squared value is equal to the maximum achievable, the ratio will be 1 and cannot (quite obviously) go past that limit. The maximum value is attained when there is only one non-zero count in each row or column. In our real Titanic dataset, V is 0.529, which (as said) can be labelled as a *strong* dependence (see also Section 4.1.6).

As we saw with ϕ, sometimes you might stumble upon references to V^2 in the literature. This value is essentially calculated as V without taking the square root. Actually, this is the original formulation by Cramér. Its interpretation parallels the one described for ϕ^2 in Section 4.1.2. I thought it prudent to give you another heads-up in case you encounter V^2 to avoid any potential confusion.

All in all, V can be thought of as measuring the association between the two cross-tabulated variables as a percentage of its possible maximum value. The fact that V takes care of the cross-tab's size makes the coefficient comparable across tables, even when they feature different sizes (a different number of rows and/or columns). Cool, eh? In 2×2 cross-tabs, the (uncorrected) ϕ coefficient and V are the same.

4.1.5 Limitations of Cramér's V

While Cramér's V offers a standardised measure of association, it is essential to understand its limitations. V performs poorly when three interrelated conditions are present: (1) concentration of observations, (2) sparse cells, and (3) uneven marginal frequencies. The first condition happens when the majority of the counts are concentrated in specific cells. The second is when some cells have very low or zero counts. The third condition is present when some rows and columns have high totals due to the concentration of counts, while others have low totals due to sparse cells. In such cases, V might overstate the strength of the association, proving pretty high even when the overall association is weak.

Let's have a look at the fictitious dataset shown in Table 4.1, where for the sake of argument we are momentarily breaking the rules by considering a 3×3 table instead of 2×2. In the table, we consider SURVIVAL and passengers' CLASS, a variable that we will be meeting again very soon in Sections 4.2.5 and 4.2.6, and in Chapter 5.

In this example, V is equal to 0.504 indicating a sizeable association in spite of the fact that the rest of the table is close to the null hypothesis of

TABLE 4.1

Cross-Tabulation of SURVIVAL Outcomes and CLASS
for a Fictitious Group of 100 Titanic Passengers

SURVIVAL	CLASS			
	1st	2nd	3rd	Total
Died	50	0	20	70
Expected	49	0.7	20.3	
Survived	0	1	1	2
Expected	1.4	0.02	0.58	
Injured	20	0	8	28
Expected	19.6	0.28	8.12	
Total	70	1	29	100

Note: Expected counts are also reported. The V coefficient is
pretty large (0.504) even if virtually all the observed counts
are very close to the expected ones. The W coefficient, on
the other hand, is 0.046, pointing to a negligible
association.

independence. In fact, all the cells (one excluded) feature an observed count
that is very close to the expected one. Only one cell (Survived–2nd Class)
features a small observed count (1) and a very small expected one (0.02). The
chi-squared value in that cell proves pretty large: $(1 - 0.02)^2 : 0.02 = 48.02$,
indicating a large difference between the observed and expected counts.
That one cell has an unduly effect on the chi-squared statistic and, in turn,
on V.

What general lesson we can learn from the above example? Many observations concentrated in specific cells can lead to large differences between
observed and expected frequencies, which in turn can inflate the chi-squared
statistic. This inflation can cause V to indicate a stronger association than
what might actually exist. By the same token, sparse cells or cells with very
low counts can also inflate the chi-squared value, especially if the expected
frequencies in these cells are also low. This is because the chi-squared statistic is sensitive to differences between observed and expected frequencies,
and even small absolute differences can lead to large relative differences
when the expected frequencies are small. Uneven marginals can result from
either a concentration of observations or from sparse cells, or both. Uneven
marginals can lead to misleading V values because the measure does not
adjust for the unevenness of the marginals, only for the size of the table. Any
one of the above conditions can cause V to be misleading. Their combined
presence can exacerbate the issue.

Quite fortunately, alternative measures have been proposed to address
the described limitation. One such measure is the W coefficient, which has
been introduced to provide a more accurate measure of the association in

situations like the ones just described. While the pretty complex details of the W coefficient are beyond the scope of this book, it is worth noting that in tables where the marginals are evenly distributed V and the W coefficient tend to offer fairly comparable results. For Table 4.1, W is equal to 0.046, indicating a negligible association. It is more in line with the fact that the cross-tab is overall very close to what we would expect under the hypothesis of independence. In our real Titanic data, on the other hand, W is equal to 0.518, which is pretty close to the calculated V.

If you are interested in knowing more about the shortcomings of V and the W alternative, you can have a look at the readings suggested in Section 7.4.5.

4.1.6 From Numbers to Meaningful Magnitudes: Interpreting Association Measures

In our exploration of chi-squared-based coefficients, we have encountered different measures that provide a numerical assessment of the strength of association. A recurring question is how to verbally articulate the magnitude of these coefficients. While numerical values provide precision, the verbal quantification of strength requires a set of guidelines for interpretation. As we have touched upon in Section 4.1.1, the guideline by Healey suggests defining a weak (small) dependence as ranging from 0.0 to 0.10, a moderate (medium) dependence from 0.11 to 0.30, and a strong (large) dependence from 0.31 to 1.0. It is important to reiterate that these categories can vary by discipline and context.

Yet another set of (admittedly tentative) thresholds has been suggested by the statistician J. Cohen. The scholar proposes a more granular interpretation, particularly when comparing associations across studies with different table sizes. Here are the thresholds for interpreting the strength of association across different df values (more on the latter soon):

df = 1: small = 0.100, medium = 0.300, large = 0.500
df = 2: small = 0.071, medium = 0.212, large = 0.354
df = 3: small = 0.058, medium = 0.173, large = 0.289
df = 4: small = 0.050, medium = 0.150, large = 0.250
df = 5: small = 0.045, medium = 0.134, large = 0.224
df = 6: small = 0.041, medium = 0.122, large = 0.204
df = 7: small = 0.038, medium = 0.113, large = 0.189
df = 8: small = 0.035, medium = 0.106, large = 0.177
df = 9: small = 0.033, medium = 0.100, large = 0.167
df = 10: small = 0.032, medium = 0.095, large = 0.158

As apparent, these thresholds are based on the df of the cross-tab. They are not to be confused with the df associated with the chi-squared test, which is equal to the number of rows minus 1 multiplied by the number of columns minus 1 (we touched upon this in Section 2.3.4). Instead, Cohen's df refers to the lesser of the number of rows or columns in the cross-tab, minus 1, and is used to scale the thresholds. For a 2×2 table, the df is equal to 1. For a 5×4 table, the df is equal to 3: the smaller number in the number of rows or columns (that is, 4) minus 1. I could continue, but I trust you got the idea.

If you are curious to know how those thresholds are devised for df larger than 1, consider that they are calculated by dividing the thresholds for 1 df by the square root of the df in question. For instance, for df = 2, the threshold for an association of small strength is equal to 0.1 (the threshold when df = 1) divided by the square root of 2 (approximately 1.41). If you do the math, 0.1 divided by 1.41 gives approximately 0.071. Similarly, the threshold for an association of medium strength is 0.3 divided by 1.41, which is approximately 0.212. The same logic applies to the other thresholds. This approach ensures a scaling of thresholds that is appropriate for the number of df.

To determine the magnitude of the association expressed by a given value of (say) V, one must first identify the appropriate df for the cross-tab under analysis. For example, if we have a 5×4 table, the df would be 3 (the smaller of the number of rows or columns, minus 1). If our calculated V is 0.175, we would compare this value to the thresholds for df = 3: 0.058 (small), 0.173 (medium), and 0.289 (large). Since 0.175 is just above 0.173, we would interpret this as an association of medium strength. The same procedure applies to the other coefficients previously discussed. However, remember that ϕ and ϕ_{corr} should only be assessed against the first row of thresholds, as they are appropriate solely for 2×2 tables.

For our real Titanic dataset (Table 2.4), V was equal to 0.529 which, for a 2×2 table (df = 1), can be interpreted as pointing to a dependence of large strength. This is consistent with the conclusion we arrive at using the thresholds touched upon earlier on at the very start of this section. We would also arrive at the same interpretation for C_{adj} and W, which were equal to 0.707 and 0.518, respectively.

It is crucial to note that the same value of a coefficient can imply different magnitudes of association depending on the cross-tab size. For instance, a V value of 0.35 in a 3×3 table (df = 2) may suggest an association of medium strength, but the same V value in a 5×5 table (df = 4) could be indicative of an association of large strength. For a table of that size, 0.35 surpasses the threshold for an association of large strength (0.25). This highlights the importance of considering the table's size when interpreting these measures, as the threshold for determining (say) a large effect becomes more accessible to reach as the cross-tab's size increases.

Lastly, a clarification is in order. In Section 4.1.4, we discussed that V is normalised for the size of the cross-tab in its denominator, which ostensibly allows for its values to be compared across tables of different sizes. This is true in the sense that V is scaled to account for table size, making it a standardised measure. However, Cohen's thresholds offer an additional layer of standardisation by providing specific benchmarks for different table sizes, thus refining the interpretation of *V* even further. In other words, while *V* is inherently comparable across table sizes, Cohen's thresholds help to contextualise what constitutes a small, medium, or large association within those sizes. This dual consideration of both the coefficient's value and the table's dimensions ensures a more nuanced interpretation of the strength of association.

4.1.7 Reflections on the Chi-Square-Based Measures of Association

Earlier sections of this chapter took us for a journey in the intricacies of chi-square-based measures of association. At first glance, everything seemed neatly packaged and straightforward. Yet, as we delved deeper, it became evident that these measures come with their own nuances and limitations. This was the reason why I had warned you to approach them with a discerning eye. In spite of their intricacies, in the realm of cross-tab analysis and statistical software, measures like C, ϕ, and Cramér's V are common. However, not all experts in the field agree on their applicability. In fact, some have argued that it is tricky to judge their magnitude because there is no operational standard one can use. Others have posited that it is difficult to meaningfully interpret those measures other than in comparative fashion (that is, larger values represent stronger associations). If you are curious about who these experts are, you can jump straight to paragraph 7.4.5, but do not forget to return; there is still much more to explore.

It is worth unpacking what those criticisms mean. I believe that those critiques are mainly referring to two related, yet distinct, aspects. First: the reference to the lack of operational standard means that there is no universally agreed-upon rulebook to interpret the scale of measures like C, ϕ, or V. For instance, what does a *V* value of 0.50 signify? Is it large or small? As we touched upon in paragraph 4.1.1 and, more extensively in paragraph 4.1.6, some rules of thumb indeed exist. Cohen's thresholds may provide a structured approach to interpretation. These guidelines, while not without their own set of limitations, offer a more granular and standardised framework for assessing the magnitude of the association, particularly across tables of varying dimensions. This does not negate the criticisms but rather acknowledges them, offering a way to navigate interpretative scenarios with some consistency. Nonetheless, the absence of agreed-upon scales makes it challenging to compare the strength of association across different studies.

The second aspect that the mentioned criticisms highlight is about the fact that, while higher coefficients indicate stronger associations, turning that into meaningful, practical, or theoretical insights is not straightforward. In other words, chi-squared-based measures of association do not have an intrinsic interpretation in terms of probability. In fact, while measures like the odds ratio, which I will cover soon, can be interpreted in terms of difference in probability between two groups (for instance, males and females) in experiencing a given outcome (dying vs. surviving), the same does not apply to the measures based on the chi-squared. By the same token, while two odds ratios can be meaningfully compared by taking their ratio (paragraph 5.2.5), the same cannot be done with, say, V. The ratio between two V (or C or ϕ) coefficients does not have a meaningful interpretation in terms of differences in probability. Due to these criticalities, some scholars gravitate towards downplaying the importance of chi-squared-based measures, or even towards dropping them altogether. They suggest using alternative tools, such as (adjusted) standardised residuals (which we have met in Section 3.1.2) or the odds ratio, which I cover in the following Sections 4.2.2, 4.2.5, and 4.2.6. These can offer clearer insights into the underlying patterns of association in a cross-tab.

4.2 Measures of Association Not Based on the Chi-Squared

4.2.1 Goodman–Kruskal's Lambda

As you surely know by now, the chi-squared test tells us if there is a relationship between two categorical variables, but it does not assume that one affects the other, and the other is affected. It treats the two variables in the same fashion, and the idea of independence that is subsumed by the test is indeed a symmetric one. To stick with our example, we know that it is likely that there is a relationship (dependence) between GENDER and SURVIVAL, and we can think of that as the former being related to the latter. The opposite holds true as well.

However, in some cases, we might have good reasons to believe that one variable (which we can call *independent variable*) affects the other (*dependent variable*). For example, we might suspect that GENDER (the independent variable) has an effect on SURVIVAL (the dependent variable).

To understand this kind of relationship, we can use measures that are not based on the chi-squared statistics and rest on totally different grounds. One of those measures is called Goodman–Kruskal's λ (pronounced *lambda*, again a Greek alphabet letter) after the last names of the two statisticians who devised it. The family of measures to which λ belongs is based on the idea of *Proportional Reduction in Error* (PRE). They assess how much knowing the independent variable (like GENDER) helps us predict the dependent variable (like SURVIVAL). It is a bit like knowing that if you see dark clouds in the sky (independent variable), you can predict it is going to rain (dependent variable). The better the prediction, the stronger the relationship between the two variables. Now, let's delve into the details of how the coefficient works.

Let's consider our Titanic data (Table 2.3), and let's consider the column variable GENDER the independent one, and the row variable SURVIVAL the dependent one. If we disregard GENDER and randomly select an observation from the total of 1309 passengers, our best guess as to its membership to either of the SURVIVAL categories would be "died" since it is the category with the largest proportion (809 out of 1309). However, this approach leads to errors, as not all the passengers belong to that category. The probability for our randomly selected observation to belong to "died" is larger (809 : 1309 = 0.62 or 62%) than the probability of belonging to the "survived" category (500 : 1309 = 0.38 or 38%). If we continuously randomly draw 1309 observations, over the long gun, 62% of them would be correct (809) while the remaining 1309 − 809 = 500 would be wrong. 500 is therefore the number of wrong guesses in predicting the membership to the categories of the dependent variable knowing nothing (that is, disregarding) about the independent variable.

Now, let's take into account the GENDER. If we randomly select a male, our best guess as to his membership to either of the dependent variable's categories would be "died" (since the latter is the category with the largest proportion: 682 out of 843). However, guessing *died* each time leads to errors, since not every male belongs to that category. But the probability for our randomly selected male to belong to that category is larger (682 : 843 = 0.81 or 81%) than the probability of belonging to the "survived" one (161 : 843 = 0.19 or 19%). If we continuously randomly draw 843 males, over the long run, 81% of our guesses would be correct (682) and the remaining 843 − 682 = 161 would be wrong. Similarly, if we randomly select a female, our best guess as to her membership to either of the dependent variable's categories would be "survived" (since the latter is the category with the largest proportion: 339 out of 466; 0.73 or 73%). Over the long run, 73% (339) of the guesses would be correct, while the remaining 466 − 339 = 127 would be wrong.

Goodman–Kruskal's λ is calculated as follows:

$$\lambda = \frac{E1 - E2}{E1}$$

where $E1$ is the number of wrong guesses in our prediction when ignoring the independent variable, and $E2$ is the total number of wrong guesses once we take into account the independent variable. In our case, $E1$ is 500, the number of wrong guesses in predicting the membership to SURVIVAL knowing nothing about GENDER. $E2$ is 288 (161 + 127), the total number of wrong guesses once we take into account the independent variable. In our dataset, λ is equal to (500 − 288) : 500 = 0.424.

How shall we interpret λ?

It expresses how much the number of wrong guesses in predicting the membership to the categories of the dependent variable is reduced by taking into account the independent variable. If the two variables are associated, there will be a given amount of reduction; the stronger the association, the larger the reduction. The value of lambda ranges from 0.0 to 1.0, inclusive. A value of 0.0 means that the variables are not associated at all, while a value of 1.0 means that the association is perfect. The value we got, 0.42, can be interpreted by saying that knowledge of GENDER improves our ability to predict

SURVIVAL by 42%. In other words, we are 42% better off knowing GENDER when attempting to predict SURVIVAL.

Goodman–Kruskal's λ offers an interesting insight into the predictive power of one variable over another. Here are a few key points to remember:

(1) Symmetry: λ is not symmetric, meaning its value can vary based on which variable is deemed independent. For example, with our Titanic dataset, designating SURVIVAL as independent and GENDER as dependent results in a λ value of 0.382. This suggests that GENDER is comparatively more informative in predicting SURVIVAL than vice versa. However, there are instances where researchers might be reluctant or unsure about assigning directionality to the association between variables. In these cases, a *symmetric* coefficient might be more fitting. The symmetric version of λ is essentially a middle ground between two asymmetric λ: one predicting A given B, and the other predicting B given A. It is like taking an average of these two measures, but giving slightly more importance (or weight) to the one that provides better information. By tweaking the PRE methodology slightly, this symmetric version of λ can be formulated. Nevertheless, I will not delve further into this symmetric version.

(2) Marginal Frequencies: λ can be misleading when the frequency of one category greatly outweighs the others. In such scenarios, when λ might register as low as 0.0 even if other association measures are significantly higher, a corrected version of λ actually exists. It will not be covered here, but it is important for you to be aware of such a scenario and the possible limitations of the coefficient. To know more about the corrected λ, you may have a look at the readings suggested in Section 7.4.5.

(3) Causality: a higher λ value denotes a stronger association, but it is crucial to understand that this does not imply a direct cause-and-effect relationship. Taking the Titanic dataset as an example: while GENDER does offer some predictive insight into SURVIVAL, it is not the definitive cause of survival. Various other elements, like a passenger's class, might play a role in this association. We will delve deeper into these intricacies in Chapter 5, where we introduce an additional variable into the cross-tab analysis.

In spite of some limitations, λ is a valuable measure that helps us quantify the strength of association between an independent and a dependent variable. It provides a more nuanced understanding of relationships between variables than symmetric measures like the chi-square-based, especially when we have a good reason to consider one variable as independent and the other as dependent. However, as with all statistical measures, it is crucial to interpret λ values in the context of the specific data and research question at hand.

4.2.2 Odds Ratio

Do not worry. I am not introducing you to gambling, and I assure you that I have never been to a casino in my life. However, odds ratio proves useful in the analysis of cross-tabs. I must admit that its underlying logic and application may prove not immediately appealing. But, on the other hand, the odds ratio (hereafter OR) turns out to be extremely useful in the analysis of what are called *stratified cross-tabs*, which we will get to know in Chapter 5. To further solidify your understanding of ORs, the Appendix provides additional hands-on examples to reinforce the concepts and their applications in additional real-world scenarios.

The OR is another way of measuring the strength of the association between pairs of categorical variables, and it is not based on the chi-squared statistic. It can be used in cross-tabs of any size. The use of OR in larger tables, while possible, becomes more complex and I will not cover that in this book. Rather, I cover a simple use of ORs in cross-tabs featuring two rows and, at least, three columns (Section 4.2.5), and in larger tables of any size (Section 4.2.6).

As the name itself says, the *odds ratio* is the ratio between two odds. Wow … that does not explain much, does it? Let me elaborate a bit more. Let's start from the individual ingredients of the OR: the *odds*.

Odds are another way of expressing the likelihood of an event to occur. The odds are defined as the ratio (division) of the number of events *that produce* an outcome to the number of events *that do not produce* the outcome. What does that mean? It is easy to describe with an example. If I roll a die, the odds of rolling 6 are 1 : 5 (0.20); there is 1 event (rolling a 6) that produces the specified outcome of 'rolling a 6' and 5 events that do not (rolling a 1, 2, 3, 4, or 5). Let's stop here with gambling, and go straight to our real Titanic dataset. Have a look at Table 4.2. The odds of a male surviving are equal to the ratio of how many males survived to how many males died: 161 : 682 = 0.24. By the same token, the odds of a female surviving are 339 : 127 = 2.67. How do we interpret those odds?

I like thinking of odds the following way, so let me repeat the odds for males while adding some extra information as I did in Table 4.2: 161 : 682 = 1 : 4.24 = 0.24. In the central part of the last bit of text, I have divided 161 by itself (it gives 1 as result), and 682 by 161 (it gives 4.24), so obtaining the ratio 1 : 4.24, which is equal to 0.24. How shall we interpret the odds 0.24 (which are less than 1)? We can focus on that 0.24 and conclude that *the odds for a male to survive are 0.24 times the odds for a male to die*. But would that make intuitive sense? Not really, at least to me. So, we can alternatively focus on that ratio 1 : 4.24 and conclude that for every male who survived (1 is in first position, which represents who survived) about 4 died. It makes intuitive sense to conclude that males are more likely to die than to survive, doesn't it?

When I say *likely* it is worth noting that I am not using it in the strict mathematical sense of probability. Throughout this book, the word likely

TABLE 4.2

Cross-Tabulation of SURVIVAL Outcomes and GENDER for the 1309 Titanic Passengers, Followed by Calculations of Survival Odds for Each Gender and the OR

SURVIVAL	GENDER		Total
	Male	Female	
Died	682	127	809
Survived	161	339	500
Total	843	466	1309

Odds surviving | Male 161 : 682 = 1 : 4.24 = 0.24
Odds surviving | Female 339 : 127 = 2.67 : 1 = 2.67
OR 2.67 : 0.24 = 11.31 : 1 = 11.31

Note: The OR indicates that females have approximately 11.31 times higher odds of surviving compared to males. In the annotation at the bottom of the table, Odds surviving | Male stands for the odds of surviving *given* that a passenger is a male. The same interpretation applies to females.

informally refers to a higher likelihood of an outcome occurring. Now that we have clarified that point, let's get back to our odds calculation. We arrive at the same conclusion if we flip things around and take the reciprocal of the odds 0.24 (1 : 0.24), which (unsurprisingly) is 4.24. Again, we conclude that for every male who survived about 4 did not; in other words, males are about 4 times more likely to die than to survive.

Let's now consider the females, and repeat what we did for the males: 339 : 127 = 2.67 : 1 = 2.67. Are you with me here? We have divided 339 by 127, and 127 by itself, so getting 2.67: 1, which is equal to 2.67. So, here we go with our interpretation of the odds for a female to survive: for every female who died (the 1 is in second position this time, and the second position represents who died) about 3 females survived. The *odds for a female to survive are 2.67 times the odds for a female to die.* Unlike males, females are more likely to survive.

Now, let's examine the OR. If we compare the *odds of a female surviving* (2.67) to the *odds of a male surviving* (0.24), it is easy to realise that females have higher odds to survival than males. Now the question is: how much higher? The OR answers that question by taking the ratio between the two odds. In our data, the OR is 2.67 : 0.24 = 11.31. This tells us that females have about 11 times higher odds to survive than males. In other words, for every male who survived, about 11 females did. Interesting, isn't it? It is something that we would not have arrived at just eyeballing the Titanic cross-tab.

Before proceeding, I need to tell you that, to come up with the OR, we do not necessarily need to preliminarily calculate the individual odds as we did. I hit the long road just for the sake of argument and to make the idea of

OR as ratio of two odds sufficiently clear (I hope I managed; did I?). However, there is a shortcut when it comes to the OR computation. To keep with our example, if we want to know the ratio of the *odds of a female surviving* to the *odds of a male surviving*, we start from the count corresponding to the cell at the intersection of the female group and the survived group (339) and multiply that by the count falling in the opposite cell along the diagonal (682, corresponding to the males who died). The result is 231,198. Then we repeat the process, starting from the count in the cell at the intersection of the male group and survived group (161), and we multiply this by the count falling in the opposite cell along the diagonal (127, corresponding to the females who died). The result is 20,447. Finally, we divide the first result by the second, which gives 11.31. If that is not clear, let's have a look at the following formula:

$$\text{Odds ratio}\left(\text{OR}\right) = \frac{\text{Females surviving } (339) \times \text{Males dying } (682)}{\text{Males surviving } (161) \times \text{Females dying } (127)} = 11.31$$

It is not by chance if the OR is also called *cross-product ratio*. In this example, at the outset, we had a clear idea as to what are our categories of interest: *females surviving* as opposed to *male surviving*. But we could have chosen *males dying* as opposed to *females dying*. Which numbers go in the numerator and which in the denominator depends on how we want to write our interpretation and on what the analytical question we want to address is. The choice is up to us.

By the way: what if there are zeros in our cross-tab? Sometimes, you might come across zeros along any of the table's diagonal. This can pose a challenge for calculating the odds ratio. When this occurs, the Haldane–Anscombe correction should be applied. This correction involves adding 0.5 to every cell of the table before computing the odds ratio.

What we have seen so far shows that we can use the OR as a measure of the strength of the dependence between our two categorical variables, GENDER and SURVIVAL in our specific case. It makes intuitive sense that, if females are 11 times more likely to survive than males, there must be a sizeable relation between falling in the female group *and* falling in the "survived" group.

Unlike many coefficients used by statisticians, the OR behaves a little bit strangely. An OR of 1 indicates the lack of a dependence (in other words, independence) between the two variables. Think of it: if for every female who survived one male survived, it means that there is no difference in the likelihood of surviving between females and males. In other words, females and males are equally likely to survive. So, an OR of 1 (as said) indicates independence. An OR different from 1 (in either direction, that is either below or above 1) indicates dependence. The larger the OR deviates from 1, the larger the strength of the dependence (association) between the two variables.

But note: below 1, the OR cannot be smaller than 0. Above 1, the OR is unbounded and can go up to infinity. Also, when one OR is the reciprocal of the other, the two ORs represent the same strength of association but in opposite directions. What does this mean? Let's revert back to our Titanic example and to the OR we have calculated earlier on. The ratio of the odds for a female to survive to the odds for a male to survive was 11.31; we concluded that females are about 11 times more likely to survive than males. So, there is an association of sizeable (let's use this generic adjective for the time being) strength between the "female" and "survived" groups. Shall we flip that 11.31 by taking its reciprocal? The reciprocal of 11.31 is 1 : 11.31 = 0.009, which can be understood as indicating that there is a negative association of sizeable (let's keep it generic) strength between the "male" and the "survived" groups.

Do you see it? When we flipped things over by taking the reciprocal of the OR, it was like making reference to the other group of the GENDER variable, that is to males this time. In the first case (OR 11.31) we were making reference to the female group; in the second case (OR 0.009) we were making reference to the other group of the GENDER variable (male). Granted: it is easy to get confused. So, my advice here is to take the proper steps (along the lines of the examples provided so far) in order to come up straight away with an OR larger than 1 (or close to 1 in case of independence). If you want to work out the OR using the *cross-product ratio* short-cut, try to have clear in your mind what are the groups of the two variables you want to make reference to. For further clarification, and to solidify your understanding of these concepts, have a look at the additional cases discussed later on in Chapter 5 and refer to the detailed examples and step-by-step interpretations provided in the Appendix.

Finally, I have to reiterate here what I said earlier when elaborating on the chi-square-based measures of association (Section 4.1.6). It is not easy to attach labels to values of association coefficients because what can be considered weak, moderate, or strong is indeed field-specific. However, some scholars have proposed tentative guidelines for the OR. The strength of the association expressed by an OR can be labelled as follows: *negligible* <2.0, *small* 2.0–3.0, *medium*: 3.0–4.0, *large*: >4.0. For our real Titanic dataset, we can conclude that there is a large (or strong) association (dependence) between GENDER and SURVIVAL, with females being strongly associated with the "survived" group. For these and additional guidelines, refer to the readings suggested in Section 7.4.5.

4.2.3 Rescaling the Odds ratio: Yule's Q

Having delved into the OR, our next stop is Yule's Q, a measure invented by the British statistician G.U. Yule that can be only used for 2 × 2 tables. While it might sound like a character from a spy novel, Q is related to the OR and provides a perhaps more intuitive interpretation of the relationship between two categorical variables. The formula for Q is:

$$Q = \frac{ad - bc}{ad + bc}$$

where, as seen before, a, b, c, and d represent the cells of a 2 × 2 cross-tab.

Q takes the difference between the products of the two diagonals (ad and bc), and divide it by their sum. If there is no clear association trend between the two variables, it means they do not have a systematic relationship. This lack of relationship is evident when the product of a and d equals that of b and c, resulting in a Q value of 0.0. A Q value approaching +1.0 suggests a strong positive association, while a value nearing −1.0 indicates a strong negative association.

That is a bit hard to figure out, isn't it?

Let's put things as follows. "Male" is a level of the GENDER variable and can also be thought of as an attribute, a quality a passenger was possessing (we covered this in Chapter 1). By the same token, being "dead" is (unfortunately) another quality, and the same applies for being "female" and having "survived". Now, let's think of our Titanic cross-tab (Table 4.2) and of each cell that is at the intersections of those attributes. If we think of the cells along the first diagonal ad, a (with its count) indicates where being "male" goes hand in hand with being "dead"; d represents where being "female" goes hand in hand with having "survived". If we move to the other diagonal, cells b and c are related to the opposite trend: b represents when being a female is associated with being "dead", whereas c represents when being a male goes hand in hand with having "survived".

The count in cell a (682) represents how many males fall in the "died" category, while the count in cell d (339) indicates how many females fall in the opposite "survived" category. Therefore, when we multiply these, we get a sense of how much the levels of GENDER tend to fall into two distinct levels of SURVIVAL. However, only taking into account one diagonal is not enough, because we actually have another diagonal and its counts to consider. The product of the counts falling in the other diagonal is representing the opposite trend: how much the levels of GENDER tend to fall in the other levels of SURVIVAL. Therefore, overall, the two diagonals are representing two different trends, two different ways in which the levels of GENDER can fall into the two levels of SURVIVAL. One trend is the opposite of the other.

As we said earlier on, Q takes the difference between the products of the two diagonals (ad and bc), and divide it by their sum. This essentially measures the difference in the strength of the two trends ($ad - bc$) relative to their combined strength ($ad + bc$). To cut the above long story short, the coefficient gives us an idea of how much bigger the ad trend is than the bc trend.

If many males are associated with being "dead" *and* many females are associated with having "survived" (the trend expressed by the ad diagonal), the counts along that diagonal will be large at the expense of the counts along the opposite diagonal, which represent the opposite trend. If that happens, there is some sizeable evidence that the two levels of GENDER are differently associated with the two levels of SURVIVAL, with males being more associated with "died" and females with "survived". Along the same lines, but in the opposite situation, we could have most of the counts concentrated in the two opposite cells along the *bc* diagonal. If those counts are large overall compared to the counts featuring the opposite *ad* diagonal, it would mean that the previous pattern still exists, but in the opposite direction (males more associated with the "survived" category; females more associated with the "died" category). If the counts featuring the two diagonals are close to each other, it means that neither trend overweighs the other. The levels of GENDER tend to haphazardly fall into the two levels of SURVIVAL (or vice versa).

For our Titanic dataset, the value of the coefficient is

$$Q = \frac{682 \times 339 - 127 \times 161}{682 \times 339 + 127 \times 161} = 0.838$$

Note that, if you have already calculated the OR for your table, then computing Q is pretty easy. For our dataset, the OR was 11.31. Q can be arrived at as follows: $(11.31 - 1) : (11.31 + 1) = 10.31 : 12.31 = 0.838$.

If we employ the guidelines (however tentative) we were mentioning in Section 4.1.6, Q would point to an association between GENDER and SURVIVAL that we could labelled as strong and positive. Why positive? It means that the bulk of our counts fall on the diagonal that represents males not surviving (*a*) and females surviving (*d*). This suggests a pattern where the majority of one group (males) experienced one outcome (not surviving), while the majority of the other group (females) experienced the opposite outcome (surviving). The outcome for males and females is largely different.

If we were to swap the positions of the columns, the bulk of the counts would shift to the other diagonal, making Q negative. This would represent a pattern where the majority of males survived and the majority of females did not survive. The negative value of Q indicates that the outcome for males and females is again largely different, but in the opposite direction. In simpler terms, the sign of Q (whether positive or negative) tells us about the dominant pattern in our 2 × 2 table.

4.2.4 Nuances and Limitations of Yule's Q

It is essential to understand the nuances and limitations of Q. When the absolute value of Q reaches 1.0, it does not always indicate an absolute association between the two variables. As you might remember (Section 4.1.1), an absolute association means that two diagonally opposite cells are zero, implying that knowing one variable gives exhaustive information about the

TABLE 4.3

(A–D) Cross-Tabulations of SURVIVAL Outcomes and GENDER for Four Fictitious
Samples of 100 Titanic Passengers

A	GENDER			B	GENDER		
SURVIVAL	Male	Female	Total	SURVIVAL	Male	Female	Total
Died	50	0	50	Died	60	20	80
Survived	0	50	50	Survived	0	20	20
Total	50	50	100	Total	60	40	100

C	GENDER			D	GENDER		
SURVIVAL	Male	Female	Total	SURVIVAL	Male	Female	Total
Died	60	10	70	Died	30	69	99
Survived	0	30	30	Survived	0	1	1
Total	60	40	100	Total	30	70	100

Note: (A) Absolute association; (B–C) complete association; and (D) negligible association. In A, both Yule's Q and ϕ achieve the maximum value of 1.0. In B, Q is equal to 1.0 and ϕ is equal to 0.612. In C, Q is equal to 1.0 and ϕ is equal to 0.802. In D, Q is equal to 1.0 and ϕ is equal to 0.06, indicating a negligible association. Remarks: (A) under absolute association, the performance of Q and ϕ is comparable; (B–C) unlike ϕ, Q does not differentiate between absolute and complete association; (D) unlike ϕ, Q achieves its maximum even when there is virtually no dependence in the cross-tab.

other. This association is portrayed in Table 4.3A, where all who died are males (and vice versa) *and* all who survived are females (and vice versa). In case of absolute association, both Q and ϕ (and V as well) achieve their maximum value of 1.0. However, as we can see from Table 4.3B–C, if any cell has a frequency of 0, Q will inevitably hit its extremes, either –1.0 or 1.0, even though the association is not absolute. This is where Q faces criticism: it does not differentiate between an absolute association and what statisticians term as *complete* association, where only one cell has a zero value. This occurs because Q is sensitive to any *unilateral* (or one-way) association, and this unilateral association achieves its maximum in both B and C. In both cases, in fact, all the males died, no matter how many females fall in the "died" and "survived" categories, and regardless of the fact that not all who died were males. Being "male" *completely* explains falling in the "died" category.

In such situations, ϕ (or V) performs differently . In fact, for Table 4.3B, ϕ is equal to 0.612, pointing to a strong association. For Table 4.3C, ϕ proves larger (0.802), reflecting a comparatively stronger association. In fact, compared to Table 4.3B, in Table 4.3C we have now 10 females who have moved from the "died" to the "survived" category, making the distribution of counts within the table a bit more symmetric (that is, with a comparatively larger proportion of counts distributed along one diagonal). However, the coefficient does not attain its maximum because there are still counts in one of the off-diagonal cells, namely the upper-right one, making the counts' distribution within the cross-tab still pretty asymmetric. Unlike ϕ, Q reaches its maximum value in both situations, and does not differentiate between the two scenarios because it reaches its maximum in both B and C where the association is, as said, complete.

Misinterpreting a Q value of 1.0 (or –1.0) as an absolute association, when it is just due to a zero frequency in a cell, can lead to incorrect conclusions. There are scenarios where the performance of Q requires more caution. In Table 4.3D, ϕ (0.06) correctly points to a negligible association, whereas Q achieves its maximum of 1.0.

The above clearly indicates that caution is advised when using Q, as this coefficient demands a more nuanced interpretation. While some scholars contend that coefficients like Q are mostly of historical interest rather than measures to be recommended in contemporary statistical practice, such conclusion could be felt as too much drastic and dismissive. If I were to give you my opinion, I believe that it is essential to recognise that every coefficient has its distinct advantages and that what could be perceived as limitations may actually reflect nuances contingent upon the way in which we conceive association, whether absolute or complete. In general, it is crucial to always provide your readers with the actual cross-tab, and to be explicit about which aspect of the association you intend to measure and its relevance to your research question(s).

4.2.5 Odds Ratios in Cross-Tabs with Two Rows and at Least Three Columns

The OR sounded pretty interesting, and it is easy to realise how immensely helpful it can prove in having a better understanding of the association pattern in our cross-tab. In larger tables (say, 5 × 4), the calculation of odds ratios is a bit complicated, and I am going to cover only a simple strategy in the next section. In tables with two rows (or columns) and three or more columns (or rows) we can still use the OR by taking the steps described in the previous section. The approach slightly changes though, and the following example will help put things into perspective. Let's stick with the Titanic dataset and cross-tabulate SURVIVAL against the passengers' CLASS.

Let's have a look at Table 4.4. What we have here is a 2 × 3 cross-tab, where there is a significant association between the two variables (chi-squared value: 128; df: 2; p-value: <0.001). How can we use odds *ratios* (note the plural) here to understand the pattern of association between SURVIVAL and CLASS? Let's picture this: we can think of considering one of the CLASS categories (aka levels) as a point of reference to which we can contrast the other two categories. In other words, we keep one category of the CLASS variable fixed and calculate the odds ratio between that category and each

TABLE 4.4

Cross-Tabulation of SURVIVAL Outcomes and CLASS for
the 1309 Titanic Passengers

	CLASS			
SURVIVAL	1st	2nd	3rd	Total
Died	123	158	528	809
Survived	200	119	181	500
Total	323	277	709	1309

remaining category. Granted, it may sound complicated, but rest assured that the following step-by-step breakdown of the procedure will help clarify things.

We have two categories pertaining to SURVIVAL (Died–Survived) and three pertaining to CLASS (1st–2nd–3rd). We decide to: (a) focus on the odds of dying; (b) choose the 1st class as the reference category, and calculate the OR between the 1st and 2nd classes, and between the 1st and 3rd classes. Let's break this down step by step.

OR for 2nd class vs. 1st class:

$$OR = (\text{Died in 2nd class} \times \text{Survived in 1st class}) : (\text{Died in 1st class} \times \text{Survived in 2nd class}) = (158 \times 200) : (123 \times 119) = 2.16$$

OR for 3rd class vs. 1st class:

$$OR = (\text{Died in 3rd class} \times \text{Survived in 1st class}) : (\text{Died in 1st class} \times \text{Survived in 3rd class}) = (528 \times 200) : (123 \times 181) = 4.74$$

Now, let's interpret the results. When it comes to the 2nd class vs. 1st class, the OR of 2.16 means that the odds of dying in the 2nd class are about 2 times higher than those of the 1st class. This indicates a higher likelihood of dying for passengers in the 2nd class compared to those in the 1st class. As for 3rd class vs. 1st class: the OR of 4.74 suggests that the odds of dying in the 3rd class are about 5 times higher than those of the 1st class. This reveals a much higher likelihood of dying for passengers in the 3rd class compared to those in the 1st class.

In conclusion, the pairwise OR comparisons reveal that the likelihood of dying increases as we move from the 1st class to the 2nd and 3rd classes (more on this in Chapter 5). The 3rd class passengers have the highest likelihood of dying among the three categories, followed by the 2nd class passengers. It sounds interesting, doesn't it? Needless to say, we could have chosen a different reference category, as well as focusing on the odds of surviving rather than dying. It is really up to us and to the specific question we seek to address.

4.2.6 Odds Ratios in Cross-Tabs of Any Size

When you are faced with a larger table, the possibilities for calculating ORs can feel like a kid in a candy store. There is a lot to choose from, as I am going to show you very soon. While complex approaches do exist (you are referred to Section 7.4.5 for further references), let me describe a simple strategy by using the fictitious dataset portrayed in Table 4.5. It is a modification of the data we encountered in the preceding section.

You might wonder, 'How many ORs can I actually calculate from this table?'. Well, consider this: for each row category, you can pair it with every other row category, and then do the same for columns. So, with three row groups, you can make three possible pairs: Died-Survived, Died-Injured, and Survived-Injured. And since we have three columns (1st, 2nd, and 3rd class), you can make three comparisons for each pair, leading to a total number of possible ORs equal to nine. That is a whopping quantity, isn't it? However, to be more precise, the actual number of *independent* ORs is smaller, and can be calculated by multiplying the number of rows minus 1 by the number of columns minus 1. In our case, $(3 - 1) \times (3 - 1) = 2 \times 2 = 4$.

What does that mean? If you calculate the OR comparing Row 1 to Row 2 and the OR comparing Row 1 to Row 3, then the OR comparing Row 2 to Row 3 can be derived as the ratio between the first two ORs. Similarly, if you know the OR comparing Column 1 to Column 2 and the OR comparing Column 1 to Column 3, then the OR comparing Column 2 to Column 3 can be derived. Therefore, given a 3×3 table, there are $(3 - 1) = 2$ independent ORs for the rows and $(3 - 1) = 2$ independent ORs for the columns. Hence, we have $2 \times 2 = 4$ independent ORs overall. The remaining five ORs can be derived from those four.

TABLE 4.5

Cross-Tabulation of SURVIVAL and CLASS for a Fictitious Group of 1579 Titanic Passengers

SURVIVAL	CLASS			
	1st	2nd	3rd	Total
Died	123	158	528	809
Survived	200	119	181	500
Injured	50	70	150	270
Total	373	347	859	1579

Note: The table breaks down the number of individuals who died, survived, or were injured within each class.

In any case, regardless of the specific number of possible ORs, we have to bear in mind that it is not just about quantity; it is about what makes sense for our research question. Let's peek at some of the ORs we can calculate from Table 4.5.

Odds of Dying (as opposed to Surviving) between 1st and 2nd classes (first two rows and first two columns):

> OR = (Died in 1st class × Survived in 2nd class) : (Died in 2nd class × Survived in 1st class) = (123 × 119) : (158 × 200) = 0.463. Passengers in the 1st class have 0.463 times the odds of dying of passengers in the 2nd class. In other words, the odds of dying in the 2nd class are about 2.16 (1 : 0.463) times higher than in the 1st class when compared to the odds of surviving.

Odds of Dying (as opposed to get Injured) between 1st and 3rd class (first and third rows, first and third columns):

> OR = (Died in 3rd class × Injured in 1st class) : (Died in 1st class × Injured in 3rd class) = (528 × 50) : (123 × 150) = 2.86. This suggests that the odds of dying in the 3rd class are 2.86 times higher than in the 1st class when compared to the odds of getting injured. In other words, passengers in the 3rd class are more likely to die than to get injured compared to passengers in the 1st class.

Odds of Surviving (as opposed to get Injured) between 2nd and 3rd class (second and third rows, second and third columns):

> OR = (Survived in 3rd class × Injured in 2nd class) : (Survived in 2nd class × Injured in 3rd class) = (181 × 70) : (119 × 150) = 0.89. This suggests that that the odds of getting injured in the 3rd class are about 1.12 (1 : 0.89) times higher than in the 2nd class when compared to the odds of surviving. To put it in another way, passengers in the 3rd class have a slightly more likely of get injured than to survive compared to those in the 2nd class.

In essence, when working with larger tables, calculating odds ratios is also about making strategic pairings. We select specific row and column categories to compare, and then compute the odds ratios to reveal the relationships between them. Each odds ratio provides a snapshot across different levels. Alternatively, we can calculate all the possible ORs. This would allow us to evaluate them in turn to spot which ones point to strong associations, and (more generally) to have a broad understanding of the patterns of association between levels. The choice is really up to us. While there are numerous possible comparisons, the key is to choose what best answers our research question(s).

4.3 Key Takeaways

- The contingency coefficient C, the ϕ coefficient, and Cramér's V measure the strength of association between categorical variables using the chi-squared statistic.

- C is derived by dividing the chi-squared value by itself plus the table's grand total, and then taking the square root. It ranges between 0.0 (indicating independence) and a maximum achievable value that is less than 1.0 and depends on the table's size..

- C_{adj} is the adjusted version of C, ensuring its value is bounded between 0.0 and 1.0, inclusive. This adjustment addresses the limitation of C having an upper ceiling that depends on the size of the cross-tab, hence allowing comparisons between tables of different sizes.

- The ϕ coefficient is the square root of the chi-squared statistic divided by the sample size.

- The corrected ϕ coefficient, ϕ_{corr}, is a normalised version of ϕ. It is calculated by dividing ϕ by ϕ_{max}, a value derived from the marginal sums of the 2×2 table. This adjusted version accounts for the fact that the uncorrected ϕ coefficient has 1.0 as maximum achievable value only under specific configurations of the marginal sums, which makes the uncorrected coefficient not directly comparable across tables with different marginals.

- Cramér's V expands upon the concept of ϕ to accommodate larger tables. It is derived from the chi-squared statistic but normalised by the sample size and the minimum dimension of the table (minus 1). This normalisation ensures the value stays between 0.0 and 1.0, and makes the coefficient comparable across tables of different sizes.

- Cramér's V is utilised for tables larger than 2×2, while the ϕ coefficient suits 2×2 tables.

- When a table features either a concentration of counts, or sparse counts, or uneven marginals (or a combination of those), Cramér's V may prove problematic; using the W coefficient as a more accurate alternative should be considered.

- Interpreting the strength of association between categorical variables requires context-specific guidelines. Some interpretation ranges define weak (small) association as 0.0–0.10, moderate (medium) association as 0.11–0.30, and strong (large) association as 0.31–1.0, though these can vary by discipline and context.

- Cohen's thresholds provide a more granular approach to interpreting the strength of association, taking into account the size of the cross-tab. For instance, a Cramér's V value of 0.35 might indicate an association of medium strength in a 3 × 3 table (df = 2) but an association of large strength in a 5 × 5 table (df = 4); this emphasises the importance of table dimensions in the interpretation process.

- Goodman–Kruskal's λ is not based on the chi-squared logic, but on the idea of Proportional Reduction in Error. It ranges from 0.0 to 1.0.

- In the Titanic example, a λ of 0.42 can be interpreted by saying that knowledge of GENDER improves our ability to predict SURVIVAL by 42%.

- The OR is another association measure. It is not reliant on the chi-squared statistic.

- An OR of 1 indicates independence between two variables. If OR > 1, there is a positive association, while OR < 1 indicates a negative association.

- The OR can theoretically range from 0 to infinity. According to tentative rules of thumb, the strength of the association indicated by the OR can be defined as: negligible (<2.0), small (2.0–3.0), medium (3.0–4.0), and large (>4.0).

- Using the Titanic dataset, it was found that females were about 11 times more likely to survive than males.

- Yule's Q is a measure related to the OR and provides an intuitive interpretation of the relationship between two categorical variables in a 2 × 2 table.

- Its value ranges between –1.0 and 1.0, with 0.0 indicating no association. A positive Q suggests a pattern where the majority of one group experienced one outcome, while the majority of the other group experienced the opposite. Conversely, a negative Q indicates the reverse pattern.

- While Yule's Q offers valuable insights, it has nuances. A Q value of 1.0 or –1.0 does not always signify an absolute association between variables. It can be influenced by zero frequencies in any cell of the 2 × 2 table, making it essential to interpret with caution.

- In cross-tabs featuring two rows and at least three columns, the OR can be calculated through pairwise comparisons.

- Using the Titanic dataset, when comparing the odds of dying for 2nd vs. 1st class passengers, the odds were about 2 times higher in the 2nd class. For 3rd vs. 1st class passengers, the odds of dying were about 5 times higher in the 3rd class.

- In large cross-tabs of any size, a simple approach for the OR calculation is to compute the ORs for all the possible 2 × 2 tables, or for some of them. The choice is up to the analyst; the key is to choose what best answers one's research question(s).

5

The Third Dimension: Adding Depth to Your Analysis

5.1 Stratified 2 × 2 Cross-Tabs

5.1.1 Introduction

I must confess that I am truly proud of you for making it this far and being exposed to crucial aspects of cross-tab analysis. I hope the journey has not been too rough. In this chapter, we take the toolkit built in previous chapters and move on to the next level, which will involve dealing with more complexity. We do not have to be afraid of complexity as long as we master the ingredients we discussed earlier on. I do believe we have all the keys to crack what is coming next.

Now, a brief interlude before we continue. While we have touched upon methods that can be applied to larger tables, our journey into a new type of cross-tab called *stratified* will primarily focus on 2 × 2 tables. Why this specific focus? Well, stratified 2 × 2 tables are a staple in many statistical analyses and offer a clear and concise framework. Larger stratified tables have definitely their place and can offer deeper insights, but they require more advanced techniques and methodologies that are beyond the immediate scope of this book. It is a bit like learning to drive; it is best to start with an automatic before shifting to a manual transmission. Nonetheless, reference to the analysis of larger tables, and to advanced methods that can be applied to them, is made in Chapter 7.

5.1.2 Partial and Marginal Tables

As you know by now, cross-tabs tabulate two categorical variables one against the other. We have been taking as an example the real Titanic dataset, where GENDER was tabulated against SURVIVAL. As you surely recall, we have been applying a number of analytical approaches (1) to assess whether there is a dependence between the two variables, (2) to test the statistical significance of the dependence, (3) to pinpoint the source(s) of association between the two variables (by taking into account the standardised chi-squared

DOI: 10.1201/9781032726328-5

residuals), and (4) to measure the strength of the dependence (by using chi-square-based measures of association, PRE measures, ORs, and Yule's *Q*).

Now, let's picture this. What if we introduce a *third variable* into the analysis? For instance, for every Titanic passenger, we do have information about the CLASS they were travelling in. We might come up with a three-way cross-tab, which would look like the 3 × 2 × 2 cross-tab presented in Table 5.1.

When we introduce a third categorical variable, we produce a table which consists of *partial tables* (aka *conditional tables*). Partial tables cross-tabulate two categorical variables (here GENDER and SURVIVAL) for each level (aka *stratum*, which is a Latin word for *layer*) of a third categorical variable (CLASS in our case). If you are wondering why partial tables are also called *conditional* tables, here is the answer. Conditional tables cross-tabulate two variables *conditional* on fixing the third variable at some level. Generally speaking, a three-way table is also called *stratified* because the very presence of the different groups (or levels, or strata) of the third variable makes the table look stratified like a cake. Yes, I am pretty hungry as I am writing this.

In Table 5.1, the first three tables from the top are the partial tables: the first cross-tabulates GENDER and SURVIVAL only for those passengers travelling in the 1st class. The second partial table does the same for passengers

TABLE 5.1

Cross-Tabulation of SURVIVAL Outcomes and GENDER, Stratified by CLASS (1st, 2nd, and 3rd), for the 1309 Titanic Passengers

CLASS	SURVIVAL	GENDER		Total
		Male	Female	
1st	Died	118	5	123
	Survived	61	139	200
	Total	179	144	323
2nd	Died	146	12	158
	Survived	25	94	119
	Total	171	106	277
3rd	Died	418	110	528
	Survived	75	106	181
	Total	493	216	709
Total	Died	682	127	809
	Survived	161	339	500
	Total	843	466	1309

Note: The first three tables represent partial (or conditional) tables for each class, breaking down the counts of individuals who died or survived, further differentiated by gender. The last cross-tab represents the marginal table and provides the totals across all classes.

from the 2nd class, and third partial table does the same for the 3rd class. The last table at the bottom, which sums all the partial tables, is called *marginal table*; it *does not* contain any information about the third variable. It is obvious to say, but let me state the obvious anyway: if you compare the marginal table to the table presented earlier on in Table 2.3, you will realise that they are the same. Unsurprisingly, in the marginal table, each cell's count is the sum of counts from the same cell location in the partial tables. In other words, taking as an example the upper-left cell of the marginal table, of those 682 males who died, 118 were travelling in the 1st class, 146 in the 2nd, and 418 in the 3rd class (118 + 146 + 418 = 682).

'Ok, ok … that's interesting', you are surely thinking, 'but why should we be willing to take into account a third variable and come up with such complex table?'. The question is spot-on; thank you for asking.

While our previous analysis has shown that GENDER and SURVIVAL are likely to be dependent and that females are more likely than males to survive (do you remember what the standardised residuals and the OR were telling us?), we could be interested in addressing another question: does the dependence between GENDER and SURVIVAL still hold when controlling for (or holding constant) the passengers' CLASS?

In other words, is there still an association between GENDER and SURVIVAL in each level of CLASS? Is the magnitude (and/or direction) of the association (if any) the same in each passenger's class? And here is another intriguing question: what if controlling for (or holding constant) a third variable not only alters the association observed in the marginal table but completely reverses it? This brings us to a fascinating and sometimes counter-intuitive statistical phenomenon known as the Simpson's Paradox. Let's dive into it, and take a short detour. Bear with me.

5.1.3 Expect the Unexpected: The Simpson's Paradox

The Simpson's Paradox is not just a quirk in statistics; it is a lesson on the importance of understanding underlying factors and the dangers of drawing conclusions from aggregated data without considering potential stratifying variables. Let's explore this paradox with two fictitious Titanic datasets, each one representing a different scenario. Let's have a look at Table 5.2.

Let's first consider the marginal table, where only GENDER and SURVIVAL are cross-tabulated. This table depicts an extreme situation, which I made up just for the sake of argument. It is pretty clear that there is indeed independence between the two variables. We do not need a formal test here: every cell in the table features the same count. Both males and females have an equal 50% chance of survival. Needless to say, the OR for this table (aka *marginal* OR) is equal to 1.

However, when we stratify our data by passenger CLASS (which, in this fictitious example, features two levels, A and B), the scenario shifts dramatically. In both partial tables, there is a significant dependence between

TABLE 5.2

Cross-Tabulation of SURVIVAL Outcomes and GENDER, Stratified by CLASS (A and B), for a Fictitious Group of 400 Titanic Passengers

		GENDER			
CLASS	SURVIVAL	Male	Female	Total	
A	Died	10	60	70	Odds Survive | Female 40 : 60 = 1 : 1.50 = 0.66
	Survived	90	40	130	Odds Survive | Male 90 : 10 = 9 : 1 = 9
	Total	100	100	200	Conditional OR 0.66 : 9 = 0.073
B	Died	90	40	130	Odds Survive | Female 60 : 40 = 1.50 : 1 = 1.50
	Survived	10	60	70	Odds Survive | Male 10 : 90 = 1 : 9 = 0.11
	Total	100	100	200	Conditional OR 1.5 : 0.11 = 13.63
Total	Died	100	100	200	Odds Survive | Female 100 : 100 = 1
	Survived	100	100	200	Odds Survive | Male 100 : 100 = 1
	Total	200	200	400	Marginal OR 1 : 1 = 1

Note: For each passenger's class, odds of survival for females and males are computed, leading to the calculation of a conditional OR that compares the gender-specific odds within that passenger's class. The marginal table provides the overall odds and OR for the entire sample (marginal OR). In the annotation to the right-hand side of the table, Odds Survive | Female stands for the odds of surviving *given* that a passenger is female. The same interpretation applies to males.

GENDER and SURVIVAL. The chi-squared statistic is 54.95 for both partial tables, with an associated p-value smaller than 0.05. The strength of the association can be labelled as strong (V is equal to 0.524 for both tables). It is worth noting that these p-values should be interpreted with caution. Section 5.2.6 elaborates on a more rigorous approach to evaluate them.

In the first partial table, the OR (aka *conditional* OR) suggests that a female in class A has 0.07 times the odds of surviving compared to a male. In other words, a male in that class is about 13.63 (1 : 0.07) times more likely to survive than a female. You can actually verify that by doing some simple math: (40 × 10) : (60 × 90) = 0.07. Contrastingly, in class B, a female is about 13.6 times more likely to survive than a male: (60 × 90) : (40 × 10) = 13.63. The counts in the partial tables are mirrored, leading to ORs that are on the reciprocal of the other. They express the same strength of association but in two opposite directions. By the way, the mirrored counts are also the reason why the chi-squared value and V for both partial tables are exactly the same.

Let's consider a second scenario, portrayed in Table 5.3. As we did before, let's consider the marginal table first. The chi-squared test indicates that there is a significant association between GENDER and SURVIVAL (chi-squared value: 4.457; df: 1; p-value: 0.035). V is equal to 0.148. The OR indicates that females are less likely to survive than males: (42 × 44) : (66 × 51) = 0.55. In other words, males are 1 : 0.55 = 1.82 times more likely to survive than females. However, if we consider the partial tables, the association

TABLE 5.3

Cross-Tabulation of SURVIVAL Outcomes and GENDER, Stratified by CLASS (A and B), for a Fictitious Group of 203 Titanic Passengers

CLASS	SURVIVAL	GENDER Male	Female	Total	
A	Died	34	8	42	Odds Survive | Female 12 : 8 = 1.50 : 1 = 1.50
	Survived	58	12	70	Odds Survive | Male 58 : 34 = 1.71 : 1 = 1.71
	Total	92	20	112	Conditional OR 1.50 : 1.71 = 0.88
B	Died	10	43	53	Odds Survive | Female 30 : 43 = 1 : 1.43 = 0.69
	Survived	8	30	38	Odds Survive | Male 8 : 10 = 1 : 1.25 = 0.80
	Total	18	73	91	Conditional OR 0.69 : 8 = 0.86
Total	Died	44	51	95	Odds Survive | Female 42 : 51 = 1 : 1.21 = 0.82
	Survived	66	42	108	Odds Survive | Male 66 : 44 = 1.50 : 1 = 1.50
	Total	110	93	203	Marginal OR 0.82 : 1.50 = 0.55

Note: On the annotations to the right-hand side of the table, see Table 5.2.

disappears. The chi-squared test for both tables proves not significant (class A, chi-squared value: 0.065; df: 1; p-value: 0.799; class B, chi-squared value: 0.066; df: 1; p-value: 0.796). It means that the odds and OR reported in Table 5.3 are measuring not significant departures from the hypothesis of independence. Not by chance they are pretty close to 1. The V coefficients point to the same conclusion: they indicate a negligible association (0.024 and 0.027 for the first and second partial table, respectively).

What can we learn from the above examples? The two fictitious scenarios showcase the intriguing nature of the *Simpson's Paradox*. In the first case, at an aggregate level (that is, in the marginal table), we observed no apparent relationship between the two variables. Still, when considering a third variable, not only a relationship emerged, but it dramatically reversed across the levels of the stratifying variable. In the second scenario, the association present at an aggregate level disappeared when considering the third variable. Such phenomena emphasise the importance of considering potential confounding variables in our cross-tab analysis. Sometimes, what is clear on the surface might mask deeper and more complex scenarios.

5.1.4 Peeling the Strata: Variable Associations through Layers

After the preceding short (and hopefully interesting) detour, let's go back to our real Titanic data, and have a look at Table 5.4.

We do know that there is a significant dependence between GENDER and SURVIVAL in the marginal table, that is in the overall Titanic dataset, where the third variable CLASS is *not* taken into consideration. We have seen this in Chapters 2 and 3. The chi-squared statistic was 365.89, with an associated p-value <0.001.

TABLE 5.4

Cross-Tabulation of SURVIVAL Outcomes and GENDER, Stratified by CLASS (1st, 2nd, and 3rd), for the 1309 Titanic Passengers

		GENDER			
CLASS	**SURVIVAL**	**Male**	**Female**	**Total**	
1st	Died	118	5	123	Odds Survive | Female 139 : 5 = 27.8 : 1 = 27.8
	Survived	61	139	200	Odds Survive | Male 61 : 118 = 1 : 1.94 = 0.52
	Total	179	144	323	Conditional OR 27.8 : 0.52 = 53.46
2nd	Died	146	12	158	Odds Survive | Female 94 : 12 = 7.8 : 1 = 7.8
	Survived	25	94	119	Odds Survive | Male 25 : 146 = 1 : 5.84 = 0.17
	Total	171	106	277	Conditional OR 7.8 : 0.17 = 45.88
3rd	Died	418	110	528	Odds Survive | Female 106 : 110 = 1 : 0.96 = 1.04
	Survived	75	106	181	Odds Survive | Male 75 : 418 = 1 : 5.57 = 0.18
	Total	493	216	709	Conditional OR 1.04 : 0.18 = 5.7
Total	Died	682	127	809	Odds Survive | Female 339 : 127 = 2.67 : 1 = 2.67
	Survived	161	339	500	Odds Survive | Male 161 : 682 = 1 : 4.24 = 0.24
	Total	843	466	1309	Marginal OR 2.67 : 0.24 = 11.13

Note: For each class, odds of survival for females and males are computed, leading to the calculation of a conditional OR that compares the gender-specific odds within that passenger's class. The last table provides the marginal totals across all classes, along with the overall odds and OR for the entire sample (marginal OR). In the annotation to the right-hand side of the table, Odds Survive | Female stands for the odds of surviving *given* that a passenger is female. The same interpretation applies to males.

To save space in Table 5.4, I did not add the result of the chi-squared test of each partial table. Just in case you are curious, the results are the following: 1st class – chi-squared statistic 131.99 with p-value <0.001 (*V*: 0.639); 2nd class – chi-squared statistic 146.46 with *p*-value <0.001 (*V*: 0.727); 3rd class – chi-squared statistic 90.58 with p-value <0.001 (*V*: 0.357). The chi-squared tests indicate that, in each partial table, there is a significant dependence between GENDER and SURVIVAL. Considering the individual *V* coefficients, each association proves pretty strong. Again, bear in mind that these *p*-values should be interpreted with caution, as Section 5.2.6 explains.

We could give a visual inspection of the table using the mosaic plot we have reviewed in Section 3.1.4. Figure 5.1 displays one mosaic plot for each level of CLASS.

It is pretty evident that, in each passenger's class, there is a discrepancy in the proportion of died and survived within each gender. Incidentally, we can see that the proportion of males increases as we move from the 1st to the 3rd class. What is more important for our analysis is that, as the class gets lower, the proportion of males dying increases, while the proportion of females surviving decreases. From the point of view of the adjusted standardised residuals, the residual associated with every table's cell proves significant

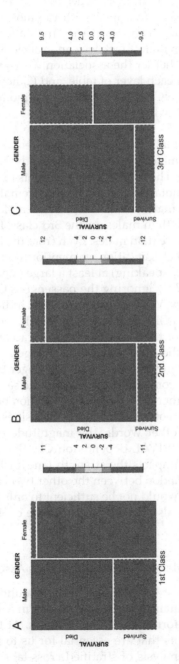

FIGURE 5.1
Mosaic plots illustrating the relationship between GENDER and SURVIVAL on the Titanic, stratified by passenger CLASS (A: 1st class; B: 2nd class; C: 3rd class). The grey-scale colour gradient reflects the adjusted standardised chi-squared residuals. Tiles with a dashed border feature negative residual. Based on data in Table 5.4.

(that is, larger than 1.96 in terms of absolute value), indicating a significant deviation from the hypothesis of independence in each cell.

The overall pattern we have described with the mosaic plot is perhaps better captured by the ORs, which summarise the patterns of association with single numbers. As a matter of fact, we can proficiently use the ORs to meaningfully examine (1) whether the association we know existing in the marginal table still exists in each layer of table, and (2) how that association behaves in each group of the CLASS variable. The annotations in Table 5.4 help us understand what story lies in each layer.

Let's have a look at the ORs for the marginal table and for each partial table. The OR for the marginal table indicates that females are 11.13 times more likely to survive than males; we know this already. Now, let's consider the OR for each partial table. The individual ingredients to work out the ORs are indicated in Table 5.4 annotations. In the 1st class, females are 53.46 times more likely to survive than males. In the 2nd class, females are about 45.88 times more likely to survive than males. In the 3rd class, females are about 5.7 times more likely to survive than males. Even from the OR perspective, if we use the tentative 3-tiered classification we mentioned in Section 4.2.2, the association proves (generally speaking) at least a large one.

What does all this tell us? By ignoring the passenger's CLASS, we would capture only part of the story. While the marginal OR indicates that there is a strong association between GENDER and SURVIVAL, the conditional ORs reveal that the magnitude of this association varies according to the passenger's class. Specifically, the conditional ORs change across the levels of CLASS: the higher the class in which a female passenger travels, the higher the likelihood of surviving (compared to a male in the same class). Overall, as the marginal OR shows, there is a sizeable association between GENDER and SURVIVAL. The strength of this association intensifies as we move from the 3rd to the 1st class. In other words, the magnitude of the association between GENDER and SURVIVAL depends on CLASS. The latter acts as a moderating variable that impacts the strength (and/or the direction, but not in this case) of the association between the other two variables. From an interpretative standpoint, it would not be sufficient to only look at GENDER and SURVIVAL; one would also need to consider the context and nuances provided by CLASS.

5.1.5 Conditional Independence and Homogeneous Association

Our conclusions at the end of the previous section sounded very interesting, at least to me. It is fascinating to see how things may change when we introduce a new piece of information in our analysis. At this point, let me introduce a few new concepts that are essential for us to know in case we will ever engage with the analysis of stratified cross-tabs. And, as usual, I promise to keep things as simple as possible.

The analysis of the stratified Titanic cross-tab that we carried out in the previous section indicates two things:

(1) That between GENDER and SURVIVAL there is *conditional depen-dence* (or, to put it the other way around, lack of conditional indepen-dence), and

(2) That there is *no homogeneous association*.

What does all this mean?

To put it in a nutshell, conditional dependence means that the association between two variables (like GENDER and SURVIVAL in our case) is *condi-tioned* on the levels of a third one (CLASS). In such a case, if we cross-tabulate the two primary variables and stratify by the third one, there is a significant association in *at least* one of the partial tables. In our case, this requisite was met in all the conditional tables. In fact, the association in each one of them proved statistically significant, as the individual chi-squared tests, the mosaic plots, and the adjusted standardised residuals were indicating, remember?

No homogeneous association means that how GENDER and SURVIVAL are associated *varies* across CLASS, the third variable we introduced. We observed that the strength of the association between GENDER and SURVIVAL was different across the levels of the CLASS variable. In fact, moving from the 3rd to the 1st class, we saw that the conditional ORs were increasing, indicating that the likelihood for a female to survive increases across the levels of CLASS. The fact that the likelihood changes across pas-sengers' class indicates that it is *not* the same across the groups (or levels, or strata) of the third variable; it is *not* homogeneous.

When there is no homogeneous association, we can also say that there is a significant *three-way association* or *interaction* among our three categorical variables. It means that how GENDER and SURVIVAL are associated (in terms of strength and direction) depends on CLASS. Conversely, if there is homogeneous association, it means that the association between the two variables (say, GENDER and SURVIVAL) is the same (in strength and direc-tion) at each level of the third variable (CLASS, in our case).

Now that we introduced the above new concepts, I need to reassure you about one thing. When I was elaborating on Table 5.4, in order to give you an idea of how the conditional ORs were changing across the levels of CLASS, I hit the long road just for the sake of argument. Of course, when we come up with a stratified cross-tab and we want to test whether the hypothesis of con-ditional independence and of homogeneous association hold, we do not have to carry out multiple chi-squared tests, calculate the conditional ORs and the marginal OR, and compare and contrast them. Of course, this can be done in a later stage to gain a deeper insight into the pattern of association. What we can preliminarily do is to carry out two formal tests, named after the statisti-cians who first devised them. I cover those tests in the next sections.

5.2 Delving Deeper: Analytical Tests and Evaluations

5.2.1 Conditional Independence and the Cochran–Mantel–Haenszel's Test

Conditional independence can be tested using the *Cochran–Mantel–Haenszel* (CMH) test. At its core, this test seeks to answer: are all the conditional ORs equal to 1? Remember, an OR of 1 suggests that the two variables are independent. You might be wondering: if the OR in each partial table is equal to 1, would this indicate independence between variables such as GENDER and SURVIVAL for each group of the CLASS variable? It would indeed.

The test provides a value named (quite fittingly) the CMH statistic. If any conditional ORs are larger or smaller than 1, the CMH statistic increases. However, if all the ORs are equal to 1, the CMH statistic becomes 0. Note that to simplify and avoid overwhelming you with mathematical jargon, I have refrained from including the complex formula for the CMH statistic. After all, understanding the concept is often more crucial than being mired in complex calculations. You can refer to the readings suggested in Section 7.4.6 if you really want to know more about the actual formula.

Not surprisingly, the CMH statistic is associated with a *p*-value. If the *p*-value is smaller than 0.05, the test indicates that the null hypothesis of conditional independence *does not* hold for our stratified table. We can conclude that *at least one* conditional (partial) table features a significant association (that is, its OR is different from 1).

But here is a key point: if associations across partial tables pull in opposite directions (that is, if some conditional ORs are smaller than 1 *and* others larger than 1), the CMH test might not work well. It could not be able to detect an existing association. This implies that a non-significant CMH statistic could either indicate no association or that no pattern of association is consistent enough to be detected. In such a scenario, it would make sense to examine the chi-squared test's result for each partial table (like we did in Sections 5.1.3 and 5.1.4). This could potentially reveal differing patterns of association that the overall CMH test might miss. We will return to this point later on, in Section 6.1.2.

If the hypothesis of conditional independence *does not* hold (there *is* conditional dependence in our stratified table), two scenarios can emerge: (1) the conditional ORs do not differ too much from one another or (2) the conditional ORs are largely different from one another.

Why are those two scenarios important? Well, the first is important because, if the conditional ORs are not too much different across the groups of the third variable, it could well happen that the differences we observe are just the result of *random variability*, which we got to know in Section 2.3.3 when we were elaborating on the variability of the chi-squared statistic. In other words, in the parent population, the conditional ORs are the same (*homogeneous*), and the differences between the conditional ORs we observe in our dataset are just because our sample is a naughty child of well-mannered parents. In this scenario, the association between the two variables across the groups of the third can be distilled down into and summarised by a single (common) OR. This first scenario is the one that we would face in case our dataset features *homogeneous association*.

On the other hand, the second scenario I was referring to would entail that the conditional ORs are really different (*heterogeneous*) from one another, and the difference between them is so large that it is unlikely that it would just be the result of random variability. In this case, the stratifying variable has an effect (in terms of strength and/or direction) on the way the other two variables are associated. It would not be wise to summarise the conditional ORs with just a common OR. The dependence between GENDER and SURVIVAL should be described class by class (that is, by each level of the CLASS variable).

Before we move on to the next section, another word of caution: the CMH test hinges on certain conditions to be effective. We already mentioned the one about the direction of associations across partial tables. Another is about sample size. While the test can handle small cell counts within partial tables, it requires (relatively) large aggregate marginal totals across all strata. In other words, while the row and column totals of the partial tables can be small, it is important that the row and column totals of the marginal table are (relatively) large. If this condition is not met, the CMH statistic might not yield a reliable measure of conditional association. It is not easy to come up with general rules of thumb for the minimum marginal sample size required by the test. Formulae do exist to come up with some figures, but they are beyond the scope of this book. If you are interested in consulting pretty technical sources, you can have a look at the readings suggested in Section 7.4.6.

5.2.2 Homogeneous Association and the Breslow–Day's Test

The hypothesis of homogeneous association can be evaluated using the *Breslow–Day* (BD) test. It aims to determine: are the conditional ORs significantly different from each other? The null hypothesis of the test is that the odds ratios do not significantly differ (are homogeneous) across strata.

Unsurprisingly, this test yields a value known as BD statistic, whose formula I refrain from including here as well. If the associated p-value is below 0.05, the test suggests to reject the null hypothesis and to conclude that the conditional ORs significantly differ from each other. In our Titanic case, such a result would indicate that how GENDER and SURVIVAL are associated *depends* on CLASS. The association should be described for each distinct passenger's class.

Conversely, if the p-value is larger than 0.05, we can accept the hypothesis of homogeneous association: the conditional ORs differ from one another just as a result of random variability. They are likely to not really differ in the parent population. In this case, we can summarise them with a common OR, which can be calculated using what is called *Mantel–Haenszel* estimate of a common odds ratio.

It should be noted that the BD test has certain limitations, particularly regarding sample size within individual strata. Unlike the CMH test, which can perform well even when some strata have small cell counts, the BD test requires a sufficiently large sample size in each stratum to ensure accurate odds ratio comparisons. This is critical because the BD test specifically assesses the homogeneity of odds ratios across different strata, and small samples in any of these can lead to unreliable results. For more insights into how sample sizes affect the BD test, and for alternative methods to test the homogeneity of ORs, refer to the readings suggested in Section 7.4.6.

5.2.3 The Mantel–Haenszel Estimate of a Common Odds Ratio

The Mantel–Haenszel (hereafter MH) estimate takes the conditional ORs and combines them into a single estimate of the overall OR. The way it does this is by giving more weight (importance) to the OR from the larger partial tables, and less weight to the OR from the smaller partial tables. This is why the MH estimate is called a *weighted* average of the ORs from each partial table.

Here is the method for averaging ORs across strata. Imagine we have two partial tables, labelled as 1 and 2; the formula for the MH estimate is:

$$OR_{MH} = \frac{\left(\dfrac{\text{TopLeft}_1 \times \text{BottomRight}_1}{\text{Total}_1}\right) + \left(\dfrac{\text{TopLeft}_2 \times \text{BottomRight}_2}{\text{Total}_2}\right)}{\left(\dfrac{\text{TopRight}_1 \times \text{BottomLeft}_1}{\text{Total}_1}\right) + \left(\dfrac{\text{TopRight}_2 \times \text{BottomLeft}_2}{\text{Total}_2}\right)}$$

That is pretty intimidating; isn't? You might be wondering why I am presenting this formula after consciously omitting the more intricate CMH and BD test statistic earlier on. The answer is simplicity. While the CMH and BD test statistic's computation can be dense and intricate, the MH estimate, despite its seemingly daunting appearance, is more straightforward by comparison, and offers a relatively easier grasp for readers keen on understanding the essence of the concept. Let me crack it for you. Essentially, what the formula is saying is that for each partial table:

(1) You calculate the product of one set of diagonals (for instance, top-left count times bottom-right count) and divide by the grand total of that partial table;

(2) You do the same for the other set of diagonals (for instance, top-right count times bottom-left count) and divide by the grand total of that partial table;

(3) Sums up all these normalised products (products divided by grand totals) from the first set of diagonals across all partial tables, which we got at step 1;

(4) Sums up all the normalised products from the second set of diagonals across all partial tables, which we got at step 2;

(5) Takes the ratio of the summed normalised products from the first set of diagonals to the summed normalised products from the second set of diagonals.

In essence, the formula is comparing the combined strength of one set of diagonals across all tables to the combined strength of the other set of diagonals. The result tells us how many times larger (or smaller) one set is compared to the other. No worries if this does not prove really clear at this point. Later on, in Section 5.2.7, I will show you how to calculate the common OR using the data from Table 5.4.

Lastly, it is important to note that the estimate of a common OR does not correspond to the marginal OR, that is to the OR calculated from the marginal table. In fact, the latter *does not* take into account the third variable (CLASS in our case), while the common OR *does* take the third variable into account because its calculation *does* involve the individual conditional ORs.

5.2.4 Beyond the Common Odds Ratio: The Cramér's *V* Alternative

If you are more inclined to use Cramér's *V* (mindful of the limitations discussed in Section 4.1.5) as a measure of association in place of the OR, there is a method to summarise this association across different strata. Now, when we have a third variable in the mix, just like with the ORs, we can calculate *V* for each level of this third variable. But how do we combine these individual scores into one overall score? This is where the concept of a weighted average comes into play again. When averaging the *V* values, we give more importance to the values from larger tables and less to those from smaller tables. This method ensures that our overall measure, the weighted average of *V*, captures the relationship between our primary variables, while also considering the influence of the third variable.

Here is the method for averaging *V* across different strata. Again, let's imagine we have two partial tables, labelled as 1 and 2; the formula for the weighted *V* is:

$$V_{\text{weighted}} = \frac{\text{Total}_1 \times V_1 + \text{Total}_2 \times V_2}{\text{Total}_1 + \text{Total}_2}$$

The formula is telling us:

(1) For each partial table, you calculate *V*, which gives you a measure of association specific to that level of the third variable; so, in our case, we come up with V_1 and V_2;

(2) The next step involves understanding the influence of each stratum: larger tables (with more observations) give us more confidence in the V value derived from them. Therefore, each V value is multiplied by the total number of observations in its respective table (Total$_1$ and Total$_2$ in our example). This process ensures that strata with more observations have a proportionally larger influence on the overall measure;

(3) After weighting each V value, you sum these weighted values together;

(4) To get the average, you then divide this sum by the total number of observations across all strata (Total$_1$ + Total$_2$); this division ensures that the final value is a true average, reflecting the overall association while considering the influence of each stratum.

This method aims to ensure that the final weighted average of V represents the overall strength of the association between the two main variables, considering the size and influence of each stratum. By doing so, we get a comprehensive measure that captures the nuances of each level of the third variable.

5.2.5 Variability in Association Across Layers: Heterogeneity and the Ratio of Conditional Ratios

Comparing the conditional ORs can help us gauging how the strength of an association varies across layers. This approach is useful for assessing the influence (if any) of the stratifying variable on the dependence between the two primary variables. In the context of our Titanic data, calculating the ratio of conditional ORs delves into how CLASS influences the relationship between GENDER and SURVIVAL. This analysis sheds light on the interaction effect, revealing how CLASS affects the relationship between the other two variables. It is important to note that this approach hinges on the prerequisite that the hypothesis of homogeneous association can be rejected (that is, the association is not homogeneous across layers).

In Sections 5.2.7 and 6.1.2, we will see the application of ratios of conditional ORs in both the Titanic example and another context. For now, it suffices to say that, in the Titanic case, a ratio of ORs for two passenger classes significantly different from 1 would indicate that the strength of the association between GENDER and SURVIVAL differs across those classes. This difference provides evidence that CLASS modulates the relationship. For example, if the ratio of the ORs between 1st and 2nd class is significantly greater than 1, it would mean that the gender disparity in survival odds is more pronounced in the 1st class compared to the 2nd class. Conversely, a ratio significantly less than 1 would indicate a gender disparity in survival odds less pronounced in the 1st class compared to the 2nd class.

We will explore the actual figures and their implications shortly, so stay tuned.

5.2.6 Fishing in Multiple Ponds: Correcting the Significance Threshold

Earlier on, when analysing Tables 5.2–5.4, we performed a chi-squared test for each individual level of CLASS to investigate if there was a dependence between GENDER and SURVIVAL in each layer of the cross-tabs. These tests provided compelling p-values, suggesting a significant association. However, I had warned you as to the importance of considering a key statistical nuance with regard to conducting multiple tests.

While the chi-squared tests for each class may seem quite conclusive, it is essential to bear in mind the potential pitfalls of running multiple tests. Imagine you are a fisherman, and you believe that there are some extraordinary rare golden fish in the waters. You decide to fish in one pond. If you catch a golden fish, you would be confident that this pond has something special. You reject the null hypothesis of *nothing special here*.

Now, let's say you fish in different ponds (passengers' classes, in our case). The chance of catching at least one golden fish by pure luck increases because you are trying multiple times. If you claim that all ponds are extraordinary based on this single catch, you might be wrong: you increased your chances by fishing more often. This means we might incorrectly conclude that there is a significant association between GENDER and SURVIVAL within each class when that might not be the case.

To safeguard against this, statisticians use various corrections for multiple comparisons, and one of the easiest to grasp is the Bonferroni correction: a special net, if you will, that sets a stricter criterion for each individual pond (or test). This special net (devised by the Italian statistician C.E. Bonferroni) ensures you are less likely to make a *false discovery* in any of the ponds you are fishing in. Mathematically speaking, if our original threshold for declaring significance is 0.05 (see Chapter 2), and we are carrying out three different tests like we did for Table 5.4, we divide 0.05 by 3. Our new threshold becomes 0.0167, making it a tougher hurdle for any individual test to be declared significant.

By applying this correction to the analysis of Table 5.4, we will only consider the associations between GENDER and SURVIVAL in individual passengers' classes as truly significant if their p-values are less (as indeed are) than the adjusted threshold of 0.0167. This extra layer of caution ensures that our findings are robust and not just a result of fishing in multiple ponds. So, even if the original p-values for each class were extremely low (<0.001), by using the corrected threshold, we significantly reduce the likelihood that we are being misled by random chance. It sound interesting, doesn't it?

5.2.7 Putting It All Together

Having introduced new concepts, let's put them to work with our real Titanic dataset, and consider the data presented in the three-way cross-tab shown in Table 5.4. We already arrived at some interesting conclusions earlier on by manually calculating the conditional and marginal ORs and making sense of them in an informal way (that is, without using any test). If we want to be more formal (you know, sometimes it is better to wear a tie to go for a work meeting rather than wearing a t-shirt and a pair of sport shoes), we can apply the CMH and the BD test in turn.

For our three-way table, the CMH statistic is 350.01, with an associated p-value smaller than 0.0001. We can distrust the hypothesis of conditional independence and conclude that there is a significant conditional dependence (at least one conditional OR is different from 1). This leads us to another intriguing question: are the conditional ORs significantly different from one another? Our BD statistic, at 43.66 with p-value less than 0.0001, helps answer this. We can reject the hypothesis of homogeneous association and conclude that there is no homogeneous association. The conditional ORs between GENDER and SURVIVAL are significantly different from one another across the levels of CLASS. GENDER and SURVIVAL are associated and (as we have seen by inspecting the conditional ORs) the magnitude of their association strengthens as we move from the 3rd to the 1st class. There is a three-way association, or interaction, among the three variables, with the stratifying variable CLASS affecting the dependence between GENDER and SURVIVAL in terms of strength (but not direction in this case). The association is to be described and evaluated passenger's class by passenger's class (that is, level by level, or stratum by stratum).

Can we quantify the role of CLASS in the association between GENDER and SURVIVAL? Let's put the ratio of conditional ORs to work and focus on the 1st and 2nd classes. The conditional OR in the 1st class is 53.46, while for the 2nd class is 45.88. Calculating the ratio of these ORs, 53.46 : 45.88, we get approximately 1.17. This suggests that, while both classes show a strong association between GENDER and SURVIVAL, the strength of this association is slightly more pronounced in the 1st class than in the 2nd. However, the contrast is stark when we compare the 1st and 3rd classes. With the 3rd class OR at 5.7, the resulting ratio of 53.46 : 5.7 (approximately 9.38) emphasises a much more pronounced (stronger) association between GENDER and SURVIVAL in the 1st class compared to the 3rd.

All in all, while GENDER and SURVIVAL are associated, the magnitude of their association is greatly influenced by CLASS. Specifically, the association between GENDER and SURVIVAL is most pronounced in the 1st class and least pronounced in the 3rd class. This suggests that the benefits (if you were a female) or disadvantages (if you were a male) associated with one's gender in terms of survival were most evident among 1st-class passengers and least evident among 3rd-class passengers. In other words, if you were in 1st class while the Titanic was sinking, you would have hoped to be a female, whereas if you were travelling in 3rd class ... well, being male or female did not really matter as much as it did in the top class. In short, by examining the ratio of conditional ORs across passengers' classes, we can obtain insights into how CLASS interacts with the relationship between GENDER and SURVIVAL.

Now, before concluding this interesting topic (it was interesting, wasn't it?), let's suppose that the BD test told us that there is a homogeneous association in our stratified Titanic cross-tab. In reality, there is no homogeneous association, but let me suppose it exists just for the sake of argument. If that were the case, we could have summarised the association across all the partial tables using the MH estimate of a common odds ratio. In our case, it would be:

$$OR_{MH} = \frac{\left(\dfrac{118 \times 139}{323}\right) + \left(\dfrac{146 \times 94}{277}\right) + \left(\dfrac{418 \times 106}{709}\right)}{\left(\dfrac{5 \times 61}{323}\right) + \left(\dfrac{12 \times 25}{277}\right) + \left(\dfrac{110 \times 75}{709}\right)} = 11.92$$

It would be interpreted as the *average odds* of surviving for a female passenger on the Titanic compared to a male passenger. In other words, *after controlling for* the passenger's class, females are *on average* 11.92 times more likely to survive than males.

What about the Cramér's V alternative? We already have the individual Vs (I reported them earlier on in Section 5.1.4), the individual grand totals and the grand total of the marginal table (if you cannot recall the latter two ingredients, have a look back to Table 5.4). The weighted average of V would be as follows:

$$V_{\text{weighted}} = \frac{323 \times 0.639 + 277 \times 0.727 + 709 \times 0.357}{1309} = 0.505$$

For clarity, 323×0.639 represents the grand total of the first partial table multiplied by the value of V for that table, and similarly for the other tables. The 1,309 at the denominator is the overall sample size. The weighted V gives us an indication of the strength of the association between GENDER and SURVIVAL while controlling for the effect of CLASS. While keeping CLASS constant, GENDER and SURVIVAL prove to be strongly associated.

As we have explored various aspects of stratified table analysis throughout this chapter, it is clear that this approach can prove quite intricate. Understanding and applying the concepts involved can be challenging. To assist in this, I have included the flowchart in Figure 5.2:

The chart serves as a visual guide through the process of the analysis of stratified 2 × 2 cross-tabs. It begins with the CMH test and, based on the outcomes, guides you through the subsequent steps. These include the application of the BD test to evaluate the homogeneity of odds ratios across strata, and, depending on the results, either to explore the interaction effects using the ratio of odds ratios or to calculate the MH common odds ratio for a summarised view of the association. The purpose of including this flowchart

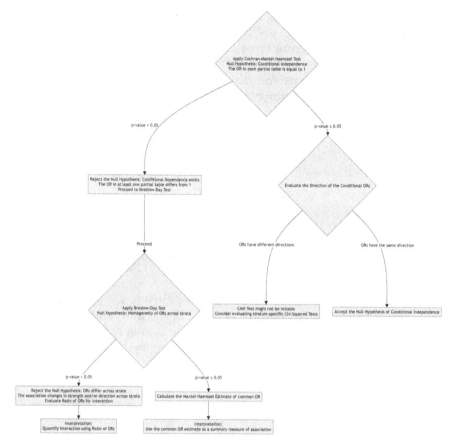

FIGURE 5.2
Decision flowchart for stratified analysis of 2 × 2 cross-tabs. The diagram outlines the sequential application of the Cochran–Mantel–Haenszel and Breslow–Day tests, and of the use of the Mantel–Haenszel estimate of a common odds ratio, guiding towards the final interpretation of the analysis results.

is to offer a visual aid that simplifies the decision-making process. It is intended as a quick reference, which encapsulates the sequence of analyses and choices in a coherent and accessible fashion.

5.3 Key Takeaways

- Stratified tables break down data into groups or *strata* based on a third variable, allowing a clearer view of how two primary variables relate within each group of the third. For the Titanic example, it shows how GENDER and SURVIVAL relate within different passenger classes.

- The Simpson's Paradox is a statistical phenomenon where an observed relationship in aggregate data disappears or reverses when data is stratified. Aggregate data can mask real associations; dissecting into subgroups can reveal the true story.

- Conditional independence implies that for any given stratum (layer) of the third variable, the two primary variables are independent.

- The Cochran–Mantel–Haenszel's (CMH) test checks if all conditional ORs are equal to 1, and thus indicates whether there is conditional independence across all strata.

- Homogeneous association implies that the relationship between the two primary variables remains consistent (in terms of strength and/or direction) across strata.

- Interaction (or three-way association) refers to the scenario where the relationship between two primary variables (like GENDER and SURVIVAL) changes depending on the level of a third variable (like CLASS). This phenomenon suggests that the association between the primary variables is not uniform (in terms of strength and/or direction) across different strata.

- In the Titanic case, the lack of homogeneous association between GENDER and SURVIVAL indicates a significant interaction, as the strength of their association varied with the passenger CLASS.

- The Breslow–Day's (BD) test assesses if the conditional ORs differ significantly from one another, giving insights into homogeneous associations.

- For the Titanic example, the CMH test indicated significant conditional dependence between GENDER and SURVIVAL, implying that at least one conditional OR is different from 1.

- The BD test indicated a lack of homogeneous association; which is to say, the association between GENDER and SURVIVAL varies across the levels of CLASS.

- The Mantel–Haenszel (MH) estimate of a common OR is insightful when the association is homogeneous. In this case, MH provides a single average OR across all strata, essentially a weighted average of individual conditional ORs.

- A weighted average of Cramér's V across strata can be calculated in order to measure the strength of the association between the two variables of interest while controlling for the effect of a third one.

- By calculating the ratios of conditional ORs, one can compare the strength of the association between the primary variables across the different levels of the third variable. This is insightful when the associations vary across strata.

- In the Titanic example, the ratio of conditional ORs between the 1st and 2nd classes was approximately 1.17, indicating a slightly stronger association between GENDER and SURVIVAL in the 1st class than the 2nd. In contrast, the ratio between the 1st and 3rd classes was approximately 9.38, highlighting a far more pronounced difference in the association.

- When performing chi-squared tests on each partial table, the risk of finding significant results merely by chance increases. The Bonferroni correction allows us to adjust our significance threshold, mitigating the risk of false positives and ensuring our conclusions are robust.

6

The Grand Finale: A Complete
Cross-Tab Analysis

6.1 Another Step-by-Step Example

6.1.1 Religiosity and Abortion Opinion

We have come a long way up to this point, and I am again really proud of you for having gone through all the preceding pages. You have been exposed to many concepts and ideas, most of which were possibly new to you. And I do hope I managed to make things easy to grasp or, at least, not so difficult to crack.

In this chapter, we analyse fictitious examples in order to put into practice what we have been reviewing so far. The example is about the result of a hypothetical survey where someone asked a random sample of 800 people their opinion about ABORTION ("favour/oppose"). For every respondent, a note was taken of their RELIGIOSITY ("not really/very religious") and of their GENDER ("female/male"). Given the general example, we will review two different scenarios, produced by two datasets featuring the same variables but yielding two different types of associations. I elaborate more on this later on, so bear with me.

Let's have a look at Table 6.1. As you know, this is a stratified cross-tab that tabulates ABORTION opinion against RELIGIOSITY for each level of GENDER. We also know that the first two tables from the top are our partial (conditional) tables; the bottom cross-tab (where GENDER is disregarded) is our marginal table. Our analytical goal is to understand (1) whether there is a dependence between ABORTION opinion and RELIGIOSITY, and (2) whether the dependence (if any) holds when controlling for GENDER; in other words, if GENDER has an effect on the association between ABORTION opinion and RELIGIOSITY.

To address the first question, we can carry out our chi-squared test on the marginal table. It suggests (chi-squared statistic: 15; df: 1, p-value: <0.001) that we can reject the hypothesis of independence, and conclude that it is likely that ABORTION opinion and RELIGIOSITY are associated in the population

DOI: 10.1201/9781032726328-6

TABLE 6.1

Cross-Tabulation of ABORTION OPINION and RELIGIOSITY, Stratified by
GENDER, for a Fictitious Group of 800 Individuals

		RELIGIOSITY			
GENDER	ABORTION OPINION	Not Really	Very Religious	Total	
Female	Favour	199	67	266	Odds Favour | NR 199 : 71 = 2.80 : 1 = 2.80
	Oppose	71	57	128	Odds Favour | VR 67 : 57 = 1.18 : 1 = 1.18
	Total	270	124	394	Conditional OR 2.8 : 1.18 = 2.38
Male	Favour	118	23	141	Odds Favour | NR 118 : 174 = 1 : 1.47 = 0.67
	Oppose	174	91	265	Odds Favour | VR 23 : 91 = 1 : 3.96 = 0.25
	Total	292	114	406	Conditional OR 0.68 : 0.25 = 2.68
Total	Favour	317	90	407	Odds Favour | NR 317 : 245 = 1.29 : 1 = 1.29
	Oppose	245	148	393	Odds Favour | VR 90 : 148 = 1 : 1.64 = 0.61
	Total	562	238	800	Marginal OR 1.29 : 0.61 = 2.13

Note: The first two tables represent partial (or conditional) tables for each gender, detailing
opinions on abortion ("Favour" or "Oppose") across levels of religiosity ("Not Really" or
"Very Religious"). For each gender, odds of being in favour for not really religious and for
very religious individuals are computed, leading to the calculation of conditional ORs.
The final table provides the marginal totals across all classes, along with the overall odds
and OR for the entire sample (marginal OR). In the annotation to the right-hand side of
the table, Odds Favour | NR stands for the odds of being in Favour *given* that one is Not
Really religious. VR stands for Very Religious.

from which we drew our sample of 800 respondents. How strong is the association? Cramér's V is equal to 0.170, indicating a dependence of moderate
(medium) strength.

We can also measure the strength of the dependence using the OR. In
this case, we can use the cross-product ratio approach we have described
in Section 4.2.2. We want to focus on the group "not really" religious (as
opposed to the "very religious") and on being in "favour" (as opposed to
"oppose"). Our marginal OR would be $(317 \times 148) : (90 \times 245) = 2.13$. People
who are not very religious are 2.13 times more likely to favour abortion
than people who are very religious. The adjusted standardised residuals
tell us that there is a significant association (a significant positive difference
between the observed and the expected counts) between being in favour and
not being really religious (adjusted standardised residual value: +4.81), and
between opposing and being very religious (+4.81).

The above results are interesting, aren't they? However, we have to bear in
mind that in our marginal table we were not taking into account GENDER.
In those 800 people we were conflating females and males. Therefore, we
may want to address the second question we put forward earlier on: does
the dependence we have found in the marginal table hold when controlling
for GENDER? Does GENDER affect (have an influence on) the association
between ABORTION opinion and RELIGIOSITY?

6.1.2 Religiosity and Abortion Opinion Controlling for Gender

The CMH test indicates that we can reject the hypothesis of conditional independence (CMH statistic: 29.73; p-value: <0.001). We conclude that there must be a significant dependence in *at least* one of the partial tables. Incidentally, we could use a mosaic plot (split by GENDER) to visually explore the association in the partial tables, as we did in Chapter 5. However, as we have seen, the ORs make a pretty good job at understanding how the association behaves across layers, and at distilling down the behaviour into single numbers. Therefore, acknowledging that visual aid could be put to work here, we keep our focus on the conditional ORs. Refer to the annotations in Table 6.1 for a breakdown of their calculation.

Along the lines of the logic we were applying earlier on when calculating the OR for the marginal table, the conditional OR for the partial table relative to females is $(199 \times 57) : (67 \times 71) = 2.38$. The conditional OR for the partial table relative to males is $(118 \times 91) : (23 \times 174) = 2.68$. Data indicate that females who are not really religious are 2.38 times more likely to favour abortion than females who are very religious. This association is significant (chi-squared: 14.99; p-value <0.001; V: 0.195). The same repeats for males: males who are not really religious are 2.68 times more likely to favour abortion than males who are very religious. This association is significant as well (chi-squared: 14.81; p-value <0.001; V: 0.191). Both chi-squared tests remain significant using the 0.025 threshold (0.05 : 2), which replaces the typical 0.05 after applying the Bonferroni correction (covered in Section 5.2.6).

We can see that the conditional ORs are close to one another. Question: are they so close just because of random variability (while in the parent population, they are likely be the same), or are they significantly different from one another (that is, they are likely to differ in the population from which we drew our sample of 800 respondents)?

The BD test indicates that we cannot reject the hypothesis of homogeneous association (BD statistic: 0.117; p-value: 0.733). In spite of the difference between the conditional ORs, we can conclude that the association between ABORTION opinion and RELIGIOSITY *does not* significantly change across the levels of the GENDER variable; the association *is homogeneous* across the levels of GENDER. In other words, GENDER *does not* affect the way in which ABORTION opinion is associated with RELIGIOSITY. Since the conditional ORs *do not statistically differ* from one another, we can summarise them using the MH estimate of a common OR, which is 2.52. Note that being a *weighted* average of the conditional ORs, the common OR is halfway between the two.

We conclude that after accounting for GENDER, data indicate that who is not really religious is, on average, 2.52 times more likely to favour abortion than who is very religious. Should we want to use V as a measure the strength of the association between ABORTION opinion and RELIGIOSITY controlling for GENDER, the weighted version of the coefficient would be

0.193. It indicates that after controlling for GENDER, ABORTION opinion and RELIGIOSITY feature an association of moderate (medium) strength.

That sounds interesting, doesn't it? Do you remember that earlier on I said that I would have provided the same example but with two different scenarios? Well, we arrived at that point. I am going to picture another scenario based on the same overall fictitious example.

Let's have a look at Table 6.2. The chi-squared test indicates that there is a significant association in the marginal table (chi-squared statistic: 4.76, df: 1, p-value: 0.029). The marginal OR is (293 × 134) : (104 × 269) = 1.40. So, if we disregard GENDER, we conclude that people who are not really religious are 1.40 times more likely to favour abortion than who is very religious. This come with no surprise; it is the same conclusion we arrived at examining the previous scenario, even though the marginal OR here is smaller than in the first scenario, indicating a comparatively weaker association.

However, let's take GENDER into account. Unlike we did before, let's for a moment postpone the CMH test and go straight to the calculation of the conditional ORs. The conditional OR for the female groups is (54 × 48) : (76 × 216) = 0.158. This indicates that females who are not really religious have 0.158 times the odds of favouring abortion than females who are very religious. Or, if we flip the OR, very religious females are 1 : 0.158 = 6.3 times more likely to favour abortion than not really religious females. I do foresee it: this is counterintuitive; how can very religious people be in favour of abortion? However, this is a fictitious scenario, do not forget. This scenario has been made up so to contrast with what is coming next with the male group. Bear with me.

Now, let's move to the male group. The conditional OR for the male groups is (239 × 86) : (53 × 28) = 13.85. This indicates that males who are not really

TABLE 6.2

Cross-Tabulation of ABORTION OPINION and RELIGIOSITY, Stratified by GENDER, for a Fictitious Group of 800

		RELIGIOSITY			
GENDER	ABORTION OPINION	Not Really	Very Religious	Total	
Female	Favour	54	76	130	Odds Favour | NR 54 : 216 = 1 : 4 = 0.25
	Oppose	216	48	264	Odds Favour | VR 76 : 48 = 1.58 : 1 = 1.58
	Total	270	124	394	Conditional OR 0.5 : 1.58 = 0.158
Male	Favour	239	28	267	Odds Favour | NR 239 : 53 = 4.51 : 1 = 4.51
	Oppose	53	86	139	Odds Favour | VR 28 : 86 = 1 : 3.07 = 0.33
	Total	292	114	406	Conditional OR 4.51 : 0.33 = 13.85
Total	Favour	293	104	397	Odds Favour | NR 293 : 269 = 1.09 : 1 = 1.09
	Oppose	269	134	403	Odds Favour | VR 104 : 134 = 1 : 1.29 = 0.78
	Total	562	238	800	Marginal OR 1.09 : 0.78 = 1.40

Note: For details, see Table 6.1.

religious are 13.85 times more likely to favour abortion than males who are very religious. Do you see what we came up with here? The two conditional ORs are indicating that the direction of the association between ABORTION opinion and RELIGIOSITY changes across the groups of the GENDER variable. For the female group, the OR is smaller than 1 (0.158) whereas, for the male group, it is larger than 1 (13.85). Data indicate that females who are not really religious have lower odds of being in favour of abortion, whereas males who are not really religious have higher odds of being in favour. It is pretty clear that the direction of the association changes across GENDER. This is different from what we observed in our earlier scenario, where the association between ABORTION opinion and RELIGIOSITY was featuring the same direction across GENDER (both the conditional ORs were in fact larger than 1).

It is apparent that, in this second scenario, we can reject the hypothesis of homogenous association. The BD test clearly indicates that (BD statistic: 181.33; p-value: <0.001). If ever we wanted a formal confirmation, the test tells us that the conditional ORs are significantly different from one another and that the hypothesis of homogeneity *does not* hold for our data. The association between ABORTION opinion and RELIGIOSITY *does* change across the levels of GENDER, and the conditional ORs *cannot* be summarised in a common OR. GENDER *does* exert an influence (this time in terms of direction) on the way in which ABORTION opinion and RELIGIOSITY are associated. Their association has to be evaluated and described gender by gender.

The ratio of the conditional ORs for males and females is 13.85 : 0.158, which is equal to 87.72. This indicates that the strength of the association between being in favour of abortion and religiosity for males is roughly 87.72 times stronger than it is for females. In other words, the association between RELIGIOSITY and ABORTION OPINION is way more pronounced for males than for females. The ratio of conditional ORs quantifies the interaction between GENDER and the relationship (association) between RELIGIOSITY and ABORTION OPINION. It gives us insights into how GENDER differentially modulates the relationship between the two primary variables.

Finally, do you remember that I asked you to postpone for a while carrying out the CMH test to formally test for conditional independence? Well, the test proves not significant (CMH statistic: 3.78, p-value: 0.062) indicating that the hypothesis of conditional independence holds. But, in this case, we have to be *careful in accepting* the results of the test. In fact, as I touched upon in Section 5.2.1, the test does not work well when the dependence across the partial tables is in opposite directions, that is, if some conditional ORs are smaller than one *and* other larger than one.

That is exactly the case in this second scenario. In spite of what the CMH test is telling us, the existence of a dependence in each partial table is indicated by the individual chi-squared tests, which prove significant for each layer (female group: chi-squared statistic 65.53, df: 1, p-value: <0.001; male group: chi-squared statistic 119.51, df: 1, p-value: <0.001). The p-values are still pointing to a significant association after applying the Bonferroni correction.

The association in each partial table is of strong/large strength. In fact, V for the female group is equal to 0.41; for the male group, it is equal to 0.54. We could get the same impression from the conditional ORs we calculated earlier on. In fact, both the OR for the female group (1 : 0.158 = 6.33) and for the male group (13.85) can be considered as indicating a strong/large association according to the tentative rule of thumb suggested in Section 4.2.2.

6.2 Being More Formal: From Sport Shoes to Tie

6.2.1 Reporting Cross-Tab Analysis Results

Before moving toward the conclusion of this chapter, I do believe that being more formal for a short while can prove beneficial. As you might remember from a few pages ago, I mentioned that sometimes it is more appropriate to wear a tie in a work meeting rather than a t-shirt and a pair of sports shoes. I was using that image to refer to the need of engaging with some statistical tests to formally ascertain a statistical hypothesis. Here, we are moving in the same territory. I can imagine some readers encountering the analysis of cross-tabs for the first time and (hopefully) appreciating the jargon-free language and layman's terms I have used in my descriptions. I also imagine that some of those, as time progresses and as they start being more and more acquainted with the described approaches, might start wondering how the results of a cross-tab analysis can be written down in a short report. Perhaps they might want to take the t-shirt and sport shoes off for a while, and wear a tie and a smart dress. I do believe that this also applies for those of you who are possibly already familiar with the procedures I have described (and with their underlying rationale) and want to be a bit more acquainted with some formal reporting.

6.2.2 Report of Religiosity and Abortion Opinion

What follows provides you with an example of how we might write down a short report (just few lines, do not worry) summarising the results of the analysis of the cross-tab that has been presented earlier on in Table 6.1. Let's picture this: first, we want to explore the dependence between RELIGIOSITY and ABORTION opinion in the marginal table (do you still remember what a marginal table is?). So, we are assuming that the marginal table is all that we have. Let's dive in an example of a formal report:

> A chi-squared test was conducted to investigate the association between the RELIGIOSITY and the ABORTION opinion. A 5% level of significance was used to evaluate significance of association, that is, the p-value was considered significant if it was less than 0.05. There was a

significant association between RELIGIOSITY and ABORTION opinion (chi-squared statistic: 15; df: 1; p-value: <0.001) with the *Not really religious* group having 2.13 times the odds of being in favour of abortion than the *Very religious* group. Cramér's V (0.170) points to a dependence of moderate strength.

6.2.3 Report of Religiosity and Abortion Opinion Controlling for Gender

If you liked the previous formal report, you will enjoy the following. Here, we are imagining that we do have information about GENDER, and that we want to analyse the relation between RELIGIOSITY and ABORTION opinion, stratified by GENDER. Again, let's dive in another example:

Stratified analysis was conducted to investigate the association between the RELIGIOSITY and ABORTION opinion after stratifying by GENDER. Conditional independence was tested using the Cochran–Mantel–Haenszel test. Homogeneity between the conditional odds ratios was tested using the Breslow–Day test. If the odds ratios were not significantly different, they were pooled to calculate Mantel–Haenszel estimate for a common odds ratio. A 5% level of significance was used to evaluate significance of associations, that is, the p-values less than 0.05 were considered significant. The CMH test indicates that the hypothesis of conditional independence can be rejected (CMH statistic: 29.73; p-value: <0.001); there is therefore a significant association in at least one of the partial tables. RELIGIOSITY proves significantly associated with ABORTION opinion for both the levels of GENDER (females: odds ratio 2.38; chi-squared p-value: <0.001; males: odds ratio 2.68; chi-squared p-value: <0.001). The BD test indicates that the hypothesis of homogeneity of the odds ratio across GENDER can be accepted (BD statistic: 0.117; p-value: 0.733). The conditional odds ratios were pooled to calculate the MH estimate of a common odds ratio, which is equal to 2.52. It suggests that the *Not Really religious* group has 2.52 times the odds of favouring abortion compared to the *Very religious* group, after adjusting (or controlling) for GENDER.

That sounds neat and clean as well, doesn't it? Well, that is all when it comes to formally reporting our tests' results. Let's take our blazer and tie off, and get more comfortable before moving on to the very end of this journey. I hope you enjoyed the ride so far.

6.3 Key Takeaways

- In the first given example, the association between RELIGIOSITY and ABORTION OPINION is analysed using a chi-squared test.

- At a 0.05 significance level, a significant association was found, with the "not really religious" group having 2.13 times the odds of favouring abortion than the "very religious" group.
- Stratified analysis introduced GENDER as a third variable to see how it influences the main relationship between RELIGIOSITY and ABORTION OPINION.
- The CMH test was used to assess conditional independence and the BD test to assess homogeneity of association.
- A significant association was observed between RELIGIOSITY and ABORTION OPINION for both genders.
- The ORs for females and males were 2.38 and 2.68, respectively.
- Homogeneity of OR across GENDER was accepted, leading to a pooled OR of 2.52.
- The second example revealed strikingly different conditional ORs between females (0.158) and males (13.85), indicating divergent associations between RELIGIOSITY and ABORTION OPINION across GENDER.
- The new scenario revealed a notable flip in the direction of the association, with "not really religious" females being less likely and "not really religious" males being more likely to favour abortion.
- The new example, as expected, rejected the hypothesis of homogeneous association as indicated by the BD test.
- The ratio of the conditional ORs indicated that the strength of the association between being in favour of abortion and religiosity for males is roughly 87.72 times stronger than it is for females.
- The second scenario underscored a cautious approach to interpreting the CMH test results due to the opposing directions of dependency, which was not a point of concern in the first example.

7

Your Next Steps in Cross-Tab Mastery

7.1 Expanding Horizon: The World beyond 2 × 2 Tables

As we wrap up our discussion on the basics of cross-tabulation analysis, it is important to understand that there is much more to explore about the subject. While our discourse has been rooted in the analysis of 2 × 2 tables, research often requires us to work with larger and more complex cross-tabs. As the number of levels of each of the two variables gets larger, you may end up dealing with tables featuring any number of rows and columns. Most of the methodologies and principles elucidated in this book, though tailored for simpler tables, are foundational and can be applied to more complex datasets. However, with larger tables, or with stratified tables with multiple layers, advanced methods are bound to become essential.

Let's stick for a while to the approaches delineated throughout this book, and let's assume we are to analyse (say) a 6 × 8 cross-tab. What can we do? Nothing prevents us from performing the chi-squared test to (preliminarily) ascertain whether an association exists between the two variables of interest. In line with what we have been discussing in Chapter 2, the test (with its associated p-value) would inform us about how likely it is that the variables are independent in the parent population. In case of small expected frequencies, the test's p-value can be obtained using the strategies previously discussed, like the $(N-1)/N$ correction, the Fisher's test, the permutation- or the Monte Carlo-based chi-squared test (Section 3.4).

By the time you read this section, you should be sophisticated enough to know that one thing is statistical significance, another is the substantive or practical importance of the association. As a consequence, even if we framed those in the context of 2 × 2 tables, some reviewed measures of association would still be relevant in the analysis of larger tables. The contingency coefficient C, Cramér's V, or W can be put to work to have a formal measure of the strength of the association (Sections 4.1.1, 4.1.4–4.1.5), provided that a significant one actually exists. Of course, should we have substantive bases to consider one of the variables as dependent on the other, we could use Goodman–Kruskal's λ to measure the extent to which the independent variable helps us predict the dependent one (Section 4.2.1). We could also employ the ORs to understand the pattern of association between pairs of levels (Section 4.2.6).

DOI: 10.1201/9781032726328-7

The chi-squared test and the use of different association measures are not the only strategies that we can meaningfully apply to both 2 × 2 and larger tables. We can inspect the chi-squared residuals in large tables as well, as we did in Section 3.1; it would help us pinpoint which cells are actually featuring a significant departure from the hypothesis of independence. This is perfectly legitimate to do, and it would actually be an essential step to take since (other things being equal) chances are that in large tables not all the cells differ from what we would expect by chance alone. If we are more inclined towards visual representations, we could use a mosaic plot as an alternative to the table of residuals (Section 3.1.4).

What I have elucidated so far sounds good, doesn't it? We can transfer our knowledge and skills from the analysis of 2 × 2 tables to the study of larger cross-tabs, and that is a big plus. However, there is a sort of tipping point we have to be aware of.

Let's imagine we have a 12 × 12 cross-tab, and we are to engage with the analysis of the chi-squared residuals or to calculate all the possible ORs. In the first case, our eyes would have to bounce between 144 cells, and for every and each residual we should locate the one(s) that proves significant. By the same token, in the second scenario, we would have to evaluate the same number of ORs (or 121 if we only consider the independent ones). That would start being a bit impractical, wouldn't it? Please note that I said impractical, not impossible. Surely, we could engage with such approaches, but it would possibly prove time-consuming for us and not so easy to digest for whoever will be our eventual reader. Also, and this is not of secondary importance, in the literature I have read so far, I never came across the analysis of stratified tables larger than 2 × 2. Indeed, the CMH test can be generalised to larger tables, but the BD test cannot.

The point I am trying to make is simple. When cross-tabs start being pretty large, other methods should be considered as a more viable and sounder option. Note that it is not just a matter of size in terms of sheer number of levels of each of the two variables. It is also a matter of the number of variables we take into account when we stratify our data. Let's consider Table 7.1 where, on the basis of the real Titanic data, we cross-tabulate GENDER and SURVIVAL (this is nothing new, right?), stratified by CLASS (again, nothing new) and (here the complication comes) by EMBARKATION PORT.

To keep the table simple, I have omitted one of the three embarkation ports and the marginal table. Table 7.1 is a four-way cross-tab (2 × 3 × 2 × 2) and, as you can see, even though the number of levels within GENDER and SURVIVAL individually considered is not that large (two levels each), once we stratify by the other two variables, the resulting table looks pretty complex. Its analysis and interpretation are bound to be cumbersome. As said, other approaches might be considered. In what follows, I limit myself to touch upon some advanced methods, and you can find further essential references in Section 7.4.7.

TABLE 7.1

Cross-Tabulation of SURVIVAL Outcomes and GENDER among 1,184 Titanic Passengers, Stratified by CLASS and EMBARKATION PORT

EMBARKATION PORT	CLASS	SURVIVAL	GENDER		Total
			Male	Female	
Cherbourg	1st	Died	42	2	44
		Survived	28	69	97
		Total	70	71	141
	2nd	Died	12	0	12
		Survived	5	11	16
		Total	17	11	28
	3rd	Died	55	9	64
		Survived	15	22	37
		Total	70	31	101
	Total	Died	109	11	120
		Survived	48	102	150
		Total	157	113	270
Southampton	1st	Died	75	3	78
		Survived	33	66	99
		Total	108	69	177
	2nd	Died	129	12	141
		Survived	20	81	101
		Total	149	93	242
	3rd	Died	313	78	391
		Survived	53	51	104
		Total	366	129	495
	Total	Died	517	93	610
		Survived	106	198	304
		Total	623	291	914

Note: For simplicity, one embarkation port (Queenstown) and the marginal table have been omitted.

7.2 When Things Get Complex: Advanced Analytical Techniques

7.2.1 Unveiling Intricacies with Log-Linear Modelling

If one aims to delve deeper into the relationship between many categorical variables, *log-linear modelling* might be a method of choice. This technique is particularly useful for understanding how different variables influence each other in multi-way tables. Imagine examining the relationship between the EMBARKATION PORT, CLASS, GENDER, and SURVIVAL on the Titanic, as shown in Table 7.1. A log-linear model could help in understanding the associations and combined influences between these categorical variables, aiding in unveiling the complex survival patterns across different groups.

For instance, let's imagine a different scenario: a study examining the relationship between AGE group ("young", "middle-aged", "elderly"), GENDER ("male", "female"), and PREFERENCE for a new product ("like", "dislike"). A log-linear model can help determine if there is an interplay between AGE and GENDER in their association with product PREFERENCE, beyond the primary effects of each variable. In simpler terms, it helps us understand whether and how AGE and GENDER together impact product PREFERENCE in a way that is not immediately apparent. For example, young females might have a different preference than expected based on the separate influences of their AGE and GENDER.

While log-linear modelling is incredibly useful for understanding the complex interplay between variables in a multi-way table, it does not provide a straightforward, easily interpretable, visualisation of the data. Especially when dealing with large tables, or when we seek to provide a visual exploration to accompany our statistical findings, another technique might be used.

7.2.2 Visualising Associations with Correspondence Analysis

This is where *correspondence analysis* comes into play. By and large, it is a technique that provides a graphical representation of cross-tabs. It is particularly useful for visualising the relationships between rows and columns in large tables of various dimensions (two-way, three-way, as well as other types of cross-tabulated data). In essence, correspondence analysis transforms the association structure present in a table into something visual that can be plotted on a chart. The graph will feature two sets of points, representing the rows and columns, respectively. The proximity of row and column points can be interpreted as indicating a variable degree of association between levels of the variables under analysis. For instance, if two levels from a row variable are close to two levels from a column variable, it suggests an association between those groups.

For a practical example, consider a study examining dietary preferences across different age groups. Imagine a large cross-tab showing various AGE GROUPS ("young", "middle-aged", "elderly") against a range of DIETARY PREFERENCES ("vegetarian", "vegan", "omnivore", "pescatarian"). Correspondence analysis can help visually represent this data, highlighting, for instance, an association between the "young" age group and "vegan" dietary preferences compared to other combinations.

For another illustrative example, consider a retail study analysing SHOPPING HABITS ("frequent", "occasional", "rare") in relation to PRODUCT CATEGORIES ("clothing", "electronics", "groceries", "home goods"). Correspondence analysis here would graphically depict the relationship between different shopping frequencies and product categories.

This analysis could reveal, for example, a proximity between "frequent" shoppers and "clothing", suggesting an association between these customers and frequent clothing purchases. Conversely, "rare" shoppers might be closely associated with "electronics", indicating that this group tends to shop infrequently, but when they do, it is more likely for electronics.

These examples underscore the power of correspondence analysis in transforming complex data into easy-to-understand visual patterns, making it an invaluable tool in diverse research fields. Inspecting a correspondence analysis scatterplot may definitely prove easier and more informative than scrutinising a large table of chi-squared residuals. It is worth noting that, in its basic version, correspondence analysis rests on the chi-squared metric, and can be thought as representing in a visual fashion the deviations from independence featuring the cross-tab under analysis. This is akin to graphically depicting the chi-squared standardised residuals. It follows that the notions covered in earlier sections of this book (especially, Sections 2.2 and 3.1) may lay the foundations to effectively grasp the underlying logic of correspondence analysis in its simplest form.

Please bear in mind that this is a very short layman description of the technique, which does not make justice to its elegant mathematical foundations, to the different interpretational aspects and nuances, to its versatility, and to the different variants that actually exist. For instance, correspondence analysis can also be used to explore three-way associations as well as cross-tabs featuring ordered categories, or cross-tabs where the dependence between the two categorical variables under analysis has a direction. That is, we might assume a causal relationship where one variable potentially influences the other, as touched upon Section 4.2.1 when we introduced the PRE measures of association. There is much more to say about the method, and I warmly suggest you to have a look at the readings provided in Section 7.4.7.

7.2.3 Employing Logistic Regression for Predictive Analysis

In certain scenarios, especially when we wish to predict the outcome of a categorical dependent variable based on one or more independent variables (aka *predictors*), *logistic regression* might be more apt. It is used to predict the outcome of a categorical dependent variable, based on one or more independent variables, which can also be categorical. The dependent variable can feature two ("yes/no", "true/false", "dead/alive") or more levels. The goal of logistic regression is to describe the relationship between the dependent variable and the predictors, enabling the estimation of the probability of a particular outcome based on the given predictors.

Consider the dataset in Table 7.1. We could use logistic regression to predict the probability of SURVIVAL based on the other categorical variables. The method could help us understand whether the effect of CLASS on SURVIVAL is influenced by GENDER, revealing, for instance, whether the impact of being in the 1st class as opposed to the 3rd class was different for men and women. Moreover, we could quantify how much the odds of SURVIVAL increase (or decrease) when moving from one passenger CLASS to another, controlling for GENDER *and* EMBARKATION PORT at the same time. This comprehensive analysis can not only illuminate the independent effects of each variable, but also explore possible interactions between predictors, providing a nuanced understanding of the factors influencing survival on the Titanic. Logistic regression can therefore be used for stratified tables analysis when the stratified cross-tab is larger than 2 × 2 and/or when we have more than one stratifying variable.

If you wish to see logistic regression framed in yet another scenario, consider a survey conducted in multiple cities: "city A", "city B", and "city C". In each city, data is collected on three AGE groups ("young", "middle-aged", "elderly"), three INCOME levels ("low", "medium", "high"), and their WILLINGNESS to purchase a luxury car ("yes/no"). Envision a cross-tab where WILLINGNESS is placed in the columns and INCOME is organised in the rows. This table is then layered by two additional variables: CITY and AGE group, creating a pretty complex cross-tab stratified by two variables (similarly to what we saw in Table 7.1). Logistic regression not only helps predict the probability of a person, based on their specific age group and income level, buying a luxury car. It would also quantify the likelihood of purchasing using ORs. It could reveal, for example, that middle-aged individuals in City B are three times as likely to purchase a luxury car compared to young individuals in the same city, assuming equal income. Furthermore, it would isolate the influence of each variable, enabling understanding how each one (for example, AGE or INCOME) independently affects the purchasing decision, while simultaneously allowing for a comparative analysis across different cities and age groups.

In the interpretation of the logistic regression results, the ORs play an important role. Their coverage in different sections of this book (Sections 4.2.2, 4.2.5–4.2.6, 5.1–5.2, and Appendix) may help establish some foundational concepts to proficiently grasp this more advanced method.

7.3 Concluding Thoughts on Advanced Analytical Techniques

Just a little disclaimer before we move on towards the end of this book: we have merely skimmed the surface of the deeper waters of those advanced analytical methodologies. Consider that a gentle dip of the toes into the ocean of complexity that these methods can navigate. Should this book find its way into enough hands (and minds!), a second expanded edition might dive deeper into some of those analytical tides; who knows.

The preceding sections underscored a pivotal lesson: the necessity of adaptable analytical methodologies when navigating through categorical data, especially as complexity increases. When data structures expand beyond 2 × 2 tables, incorporating advanced methods such as log-linear modelling, correspondence analysis, or logistic regression becomes paramount to accurately understand, represent, or predict patterns within the data.

However, it is important to note that while these advanced techniques provide enhanced analytical capabilities for more complex datasets, they each come with their own nuances, requirements, and underlying assumptions. For instance, logistic regression requires designating one variable as dependent and the others as independent. We cannot take for granted that one can always make such obvious causal ordering.

In conclusion, the choice of analytical strategy is fundamental, whether you are dissecting intricate, multi-layered cross-tabs or attempting to visualise or predict outcomes from categorical variables. Foundational principles and methodologies continue to hold value, but their practicality may diminish with increasing data complexity. This underscores the importance of methodological advancement in step with escalating analytical demands, a theme that we might explore further in potential future editions of this work.

7.4 The Explorer's Toolkit: Recommended Readings and Resources

7.4.1 Introduction

As you must have noted, throughout this book, I did not provide you with any in-text citation and reference to relevant academic literature. I deliberately chose to do that because I did not want to distract you and I wanted the style to be colloquial rather than academic. However, that did not mean to convey the idea that what this book has covered is not backed up by previous literature, nor that everything I have been covering did not build on the work of other scholars. It is right the opposite.

The following sections intend to provide some essential bibliographical references that you may want to consult in case you got interested in cross-tab analysis and wish to know more. The references are organised by chapter and topic; within each topic, references are sorted in chronological order to give a sense of historical development.

The Titanic dataset I have been using throughout this book is widely available and can be accessed from various sources. The one I used can be found (as of October 2023) at the following URL: https://hbiostat.org/data/repo/titanic3.xls.

The formal reports of cross-tab analysis, which I have covered in Section 6.2, rest on the report generated by *Statulator*, a very useful online statistical tool developed by Navneet Dhand and Mehar Khatkar. It includes facilities for chi-squared and stratified cross-tab analysis. As of March 2023, the tool can be accessed from the following URL: https://statulator.com/.

As touched upon in the Introduction, the cross-tabs in this work have been produced using the *Jamovi* software, which offers a user-friendly interface for statistical analysis. The tables of standardised residuals and the mosaic plot have been produced with the *chi-square* and *vcd* R packages, respectively.

The *chi-square* package is designed to perform the chi-squared test and its $(N - 1)/N$ corrected form, and offers permutation- and Monte Carlo-based p-values. It includes various measures of association, among which the contingency coefficient C, ϕ (both un-corrected and corrected), Cramér's V, W, Goodman–Kruskal's λ, odds ratio, Yule's Q, and others. The package also provides different types of chi-squared standardised residuals. On the other hand, the *vcd* package specialises in visualising categorical data. It proves an invaluable tool as it offers a range of visualisation techniques, datasets, summaries, and inference procedures tailored for categorical data. For more information, you can visit their respective URLs:

Jamovi: https://www.jamovi.org/
chisquare: https://cran.r-project.org/package=chisquare
vcd: https://cran.r-project.org/package=vcd

7.4.2 Essential Readings for Chapter 1

If you are new to statistics and statistical concepts, I do believe that you will enjoy reading the following brilliant book. When I first approached statistics, I really liked its content and style. Actually, I borrowed from its author the idea of trying to explain things in simple terms, accessible to anyone. I do not know if I managed, but for sure he did. The book features a short section about the chi-squared test; I do believe that it can benefit you:

Rowntree, D. (1981). *Statistics without tears: An introduction for non-mathematicians.* Penguin. ISBN 9780141987491.

7.4.3 Essential Readings for Chapter 2

The following brilliant book definitely covers both basic and more advanced topics in statistics; it shares with the preceding work the engaging down-to-earth style. It contains a great section about the hypothesis of independence and the chi-squared test. I borrowed from it the idea of using a simulation of repeat sampling to illustrate the rationale of the distribution of the chi-squared statistic under the null hypothesis. What a brilliant idea! Kudos to the author:

Linneman, T. J. (2018). *Social statistics: Managing data, conducting analyses, presenting results* (3rd ed.). Routledge. ISBN 9780415805018.

The following books (including the one published by Agresti in 2007, cited later on) are a must when it comes to the analysis of categorical data in general, and of cross-tabs in particular. They also cover measures of association, both chi-squared-based and non-chi-squared based (such as Goodman–Kruskal's λ), and prove very interesting readings for those already acquainted with statistics. Note that the book by Fagerland and colleagues is an exceptionally technical and comprehensive treatment of the analysis of cross-tabulations, including the analysis of stratified cross-tabs and related tests, which we have covered in Chapter 5:

Reynolds, H. T. (1984). *Analysis of nominal data* (2nd ed.). Sage University Paper Series on Quantitative Research Methods, Vol. 7. Sage. https://doi.org/10.4135/9781412983303.

Everitt, B. S. (1992). *The analysis of contingency tables* (2nd Ed). Chapman and Hall/CRC. ISBN 9780367450410.

Sheskin, D. J. (2011). *Handbook of parametric and nonparametric statistical procedures* (5th ed.). Chapman & Hall/CRC. ISBN 9781439858011.

Fagerland, M. W., Lydersen, S., & Laake, P. (2017). *Statistical analysis of contingency tables*. CRC Press. ISBN 9781466588172.

For the use of simulations in statistics in general, and for the calculation of the Monte Carlo *p*-value for the chi-squared test, see the following works:

Agresti, A., Wackerly, D., & Boyett, J. M. (1979). Exact conditional tests for cross-classifications: Approximation of attained significance levels. *Psychometrika*, 44(1), 75–83. https://doi.org/10.1007/BF02293786.

Mooney, C. (1997). *Monte Carlo simulation*. SAGE Publications, Inc. https://doi.org/10.4135/9781412985116

Howell, D. C. (2011). *Statistical methods for psychology* (8th ed.). Wadsworth Publishing. ISBN 9781111835484.

Utts, J. M. (2014). *Seeing through statistics* (4th ed.). Brooks/Cole. ISBN 9781285050881.

Barceló, J. A. (2018). Chi-Square analysis. In *The Encyclopedia of Archaeological Sciences* (pp. 1–5). Wiley. https://doi.org/10.1002/9781119188230.saseas0090.

7.4.4 Essential Readings for Chapter 3

The following article is a must when it comes to know more about the adjusted standardised residuals:

Haberman, S. J. (1973). The analysis of residuals in cross-classified tables. *Biometrics*, 29(1). https://doi.org/10.2307/2529686.

The following is an interesting and concise guide to the analysis of cross-tabs, odds ratio, measures of association, and more; it also elaborates on the chi-squared residuals:

Sharpe, D. (2015). Chi-square test is statistically significant: Now what? *Practical Assessment, Research, and Evaluation, 20*(8), 1–10. https://doi.org/10.7275/tbfa -x148.

For the mosaic plot and other tools for visualising categorical data, see:

Friendly, M. (2000). *Visualizing categorical data*. SAS Institute. ISBN 1580256600

If you want to know more about the issue of small expected counts in the context of the chi-squared test, and on its robustness in such situation, the following pretty technical articles are worth reading; the last one also provides a useful compilation of different scholars' opinion about the smallest tolerable expected counts:

Lewontin, R. C., & Felsenstein, J. (1965). The robustness of homogeneity tests in 2 x N tables. *Biometrics, 21*(1), 19–33. https://doi.org/10.2307/2528349.

Bradley, D. R., & Cutcomb, S. (1977). Monte Carlo simulations and the chi-square test of independence. *Behavior Research Methods & Instrumentation, 9*(2), 199–202. https://doi.org/10.3758/BF03214499.

Camilli, G., & Hopkins, K. D. (1978). Applicability of chi-square to 2×2 contingency tables with small expected cell frequencies. *Psychological Bulletin, 85*(1), 163–167. https://doi.org/10.1037/0033-2909.85.1.163.

Larntz, K. (1978). Small-sample comparisons of exact levels for chi-squared goodness-of-fit statistics. *Journal of the American Statistical Association, 73*(362), 253–263.

Bradley, D. R., Bradley, T. D., McGrath, S. G., & Cutcomb, S. D. (1979). Type I error rate of the chi-square test in independence in RxC tables that have small expected frequencies. *Psychological Bulletin, 86*(6), 1290–1297. https://doi.org/10.1037/0033 -2909.86.6.1290

Berry, K. J., & Mielke, P. W., Jr. (1986). R by C chi-square analyses with small expected cell frequencies. *Educational and Psychological Measurement, 46*(1), 169–173. https://doi.org/10.1177/0013164486461018.

Ruxton, G. D., & Neuhäuser, M. (2010). Good practice in testing for an association in contingency tables. *Behavioral Ecology and Sociobiology, 64*(9), 1505–1513. https:// doi.org/10.1007/s00265-010-1014-0.

Kroonenberg, P. M., & Verbeek, A. (2018). The tale of Cochran's rule: My contingency table has so many expected values smaller than 5, What am I to do? *The American Statistician, 72*(2), 175–183. https://doi.org/10.1080/00031305.2017 .1286260.

Van Auken, R. M., & Kebschull, S. A. (2021). Type I error convergence of three hypothesis tests for small RxC contingency tables. In Y. Hu (Ed.), *RMS: Research in Mathematics & Statistics, 8*(1). https://doi.org/10.1080/27658449.2021.1934959.

For average expected frequency rule, see the following works; note that Roscoe and Byars suggest an average expected count of 5 or more, while Zar proposes a slightly more conservative figure of at least 6. Both works agree on the need of having an average expected count of at least 10 when testing at a significance level of 0.01:

Roscoe, J. T., & Byars, J. A. (1971). An investigation of the restraints with respect to sample size commonly imposed on the use of the chi-square statistic. *Journal of the American Statistical Association, 66*(336), 755–759. https://doi.org/10.1080 /01621459.1971.10482341.

Moore, D. S. (1986). Tests of chi-squared type. In R. B. D'Agostino & M. A. Stephens (Eds.), *Goodness-of-fit techniques* (pp. 63–93). Marcel Dekker, Inc. ISBN 9780367580346.

Greenwood, P. E., & Nikulin, M. S. (1996). *A guide to chi-squared testing.* John Wiley & Sons. ISBN 9780471557791.

Zar, J. H. (2014). *Biostatistical analysis* (5th ed.). Pearson New International Edition. ISBN 9781292024042.

For the $(N − 1)/N$ correction to the chi-squared statistic, see the following pretty technical articles:

Pearson, E. S. (1947). The choice of statistical tests illustrated on the interpretation of data classed in a 2 × 2 table. *Biometrika, 34*(1–2), 139–167. https://doi.org/10.1093 /biomet/34.1-2.139

Upton, G. J. G. (1982). A comparison of alternative tests for the 2 × 2 comparative trial. *Journal of the Royal Statistical Society: Series A (General), 145*(1), 86. https://doi.org /10.2307/2981423.

Rhoades, H. M., & Overall, J. E. (1982). A sample size correction for Pearson chi-square in 2×2 contingency tables. *Psychological Bulletin, 91*(2), 418–423. https:// doi.org/10.1037/0033-2909.91.2.418.

Richardson, J. T. (1994). The analysis of 2 × 1 and 2 × 2 contingency tables: An historical review. *Statistical Methods in Medical Research, 3*(2), 107–133. https://doi.org /10.1177/096228029400300202.

Campbell, I. (2007). Chi-squared and Fisher–Irwin tests of two-by-two tables with small sample recommendations. *Statistics in Medicine, 26*(19), 3661–3675. https:// doi.org/10.1002/sim.2832.

Martín Andrés, A. (2008). Comments on 'Chi-squared and Fisher–Irwin tests of two-by-two tables with small sample recommendations' by I. Campbell, Statistics in Medicine 2007; 26:3661–3675. *Statistics in Medicine, 27*(10), 1791–1795. https://doi .org/10.1002/sim.3169.

Richardson, J. T. E. (2011). The analysis of 2 × 2 contingency tables-Yet again. *Statistics in Medicine, 30*(8), 890. https://doi.org/10.1002/sim.4116.

Busing, F. M. T. A., Weaver, B., & Dubois, S. (2015). 2 × 2 Tables: a note on Campbell's recommendation. *Statistics in Medicine, 35*(8), 1354–1358. https://doi.org/10.1002 /sim.6808.

For other corrections, see:

Serra, N., Rea, T., Di Carlo, P., & Sergi, C. (2022). Continuity correction of Pearson's chi-square test in 2 × 2 contingency tables: A mini-review on recent development. *Epidemiology, Biostatistics, and Public Health, 16*(2). https://doi.org/10.2427/13059.

On the Fisher test and its limitations, see for example the following works; the first also elaborates on how to reduce the conservativeness of the test:

Agresti, A. (2007). *An Introduction to Categorical Data Analysis* (2nd ed.). Wiley. ISBN 9780471226185.

Crans, G. G., & Shuster, J. J. (2008). How conservative is Fisher's exact test? A quantitative evaluation of the two-sample comparative binomial trial. *Statistics in Medicine, 27*(18), 3598–3611. https://doi.org/10.1002/sim.3221.

Choi, L., Blume, J. D., & Dupont, W. D. (2015). Elucidating the foundations of statistical inference with 2 × 2 tables *PLoS ONE, 10*(4), e0121263. https://doi.org/10.1371/journal.pone.0121263.

For an elegant and easy-to-follow description of the permutation-based chi-squared test, see:

Agresti, A., Franklin, C., & Klingenberg, B. (2022). *Statistics: The art and science of learning from data* (5th ed.). Pearson Education. ISBN 9780136846932.

For the Monte Carlo-based chi-squared test, see the readings suggested in Section 7.4.3.

7.4.5 Essential Readings for Chapter 4

If you want to delve deeper into the measures of categorical association, the first of the following books is a very useful resource and also covers other PRE measures besides Goodman–Kruskal's λ. The second work, while being a more general treatment of measures of effect size, also covers coefficients that can be used in categorical data analysis:

Liebetrau, A. (1983). *Measures of association*. SAGE Publications, Inc. https://doi.org/10.4135/9781412984942.

Grissom, R. J., & Kim, J. J. (2012). *Effect sizes for research* (2nd ed.). Routledge. ISBN 9780415877688.

For the corrected version of the ϕ coefficient, see:

Cureton, E. E. (1959). Note on φ/φmax. *Psychometrika, 24*(1), 89–91. https://doi.org/10.1007/bf02289765.

Liu, R. (1980). A note on phi-coefficient comparison. *Research in Higher Education, 13*(1), 3–8. https://doi.org/10.1007/bf00975772.

Davenport, E. C., Jr., & El-Sanhurry, N. A. (1991). Phi/Phimax: Review and synthesis. *Educational and Psychological Measurement, 51*(4), 821–828. https://doi.org/10.1177/0013164491051004.

Rasch, D., Kubinger, K. D., & Yanagida, T. (2011). *Statistics in psychology using R and SPSS*. Wiley. ISBN 9780470971246.

For the original formulation of Cramér's *V*, see:

Cramer, H. (1946). *Mathematical methods of statistics*. Princeton University Press.

The coefficients ϕ^2 and V^2 are discussed, for example, in the following books (see also the work by Reynold cited in Section 7.4.3):

Blalock, H. M., Jr. (1960). *Social statistics.* McGraw-Hill Book Company.
McCall, G. S. (2018). *Strategies for quantitative research: Archaeology by numbers.* Routledge. ISNB 9781138632530.

If you want to know more about the limitations of Cramér's *V*, the following articles make some interesting points; notably, the second one proposes the *W* coefficient and criticises the use of Yule's *Q*:

Berry, K. J., Johnston, J. E., & Mielke, P. W., Jr. (2006). A measure of effect size for R × C contingency tables. *Psychological Reports, 99*(1), 251–256. https://doi.org/10.2466/pr0.99.1.251-256.
Kvålseth, T. O. (2018). An alternative to Cramér's coefficient of association. *Communications in Statistics - Theory and Methods, 47*(23), 5662–5674. https://doi.org/10.1080/03610926.2017.1400056.

The following is an interesting book on the use of statistics in the social sciences; it features a section about the analysis of cross-tabs. I borrowed from this author the first of the 3-tiered classifications of the strength of categorical association covered in Chapter 4; it also elaborates on Goodman–Kruskal's λ:

Healey, J. F. (2013). *The essentials of statistics: A tool for social research.* Wadsworth-Cengace Learning. ISBN 9781111829568.

If you are interested in Cohen's classification scheme, the following pretty advanced book is the way to go:

Cohen, J. (1988). *Statistical power analysis for the behavioral sciences* (2nd ed.). L. Erlbaum Associates. ISBN 9780203771587.

For a critical examination of proposed benchmarks for effect sizes and the criticism of their rigid interpretation, the following resource provides a comprehensive analysis:

Ellis, P. D. (2010). *The essential guide to effect sizes: Statistical power, meta-analysis, and the interpretation of research results.* Cambridge University Press. ISBN 9780511761676.

In Section 4.1.7, I elaborated on the dissatisfaction of some statisticians towards measures of categorical association, such as the ϕ coefficient and Cramér's *V*. I made reference to two opinions: the first is expressed in the book by Reynolds (cited in Section 7.4.3); the second is after the following interesting book, which also features a section on the calculation and interpretation of odds ratios in 2 × 3 cross-tabs:

Agresti, A., & Finlay, B. (2008). *Statistical methods for the social sciences* (4th ed.). Prentice Hall. ISBN 9780130272959.

For information about complete and absolute association, in the context of ORs and Yule's *Q*, see:

Fienberg, S. E. (2007). *The analysis of cross-classified categorical data* (2nd ed.). Springer. ISBN 9780387728247.

For the corrected version of Goodman–Kruskal's λ, see:

Kvålseth, T. O. (2018). Measuring association between nominal categorical variables: An alternative to the Goodman–Kruskal lambda. *Journal of Applied Statistics, 45*(6), 1118–1132. https://doi.org/10.1080/02664763.2017.1346066.

If you want to know more about the use of odds ratio in the analysis of categorical data, then the following are for you:

Rudas, T. (1997). *Odds ratios in the analysis of contingency tables*. SAGE Publications, Ltd. ISBN 9780761903628.
Bland, J. M. (2000). Statistics notes: The odds ratio. *BMJ, 320*, 1468–1468. https://doi .org/10.1136/bmj.320.7247.1468.
Sedgwick, P. (2013). Odds and odds ratios. *BMJ, 347*, f5067–f5067. https://doi.org/10 .1136/bmj.f5067.

On the Haldane–Anscombe correction in the calculation of odds ratios, besides the book by Pagano and Gauvreau cited in Section 7.4.6, see the following books; note that the second also covers other types of corrections:

Fleiss, J. L., Levin, B., & Paik, M. C. (2003). *Statistical methods for rates and proportions.* Wiley. ISBN 9780471526292.
Upton, G. J. G. (2017). *Categorical data analysis by example.* John Wiley & Sons. ISBN 9781119307914.

I borrowed from the first of the following (pretty advanced) articles the tentative 4-tiered classification of the strength of the association expressed by an odds ratio; however, that is not the only guideline suggested by scholars, as other ones do exist (see the second article below):

Ferguson, C. J. (2009). An effect size primer: A guide for clinicians and researchers. *Professional Psychology: Research and Practice, 40*(5), 532–538. https://doi.org/10 .1037/a0015808.
Chen, H., Cohen, P., & Chen, S. (2010). How big is a big odds ratio? Interpreting the magnitudes of odds ratios in epidemiological studies. *Communications in Statistics - Simulation and Computation, 39*(4), 860–864. https://doi.org/10.1080 /03610911003650383.

7.4.6 Essential Readings for Chapters 5 and 6

In the following books, I have found a very good and easy-to-follow treatment of the analysis of stratified cross-tabs. The first work describes the method devised by Mantel and Haenszel to test the homogeneity of ORs across strata, whereas the second elaborates on the Breslow–Day test. For the former test, which is not covered in the present work, and for other concurrent tests, have a look at the book by Lachin and at the article by Almalik and van den Heuvel cited later on. The book by Azen and Walker is an extremely useful resource on categorical data analysis in general. The topics covered are advanced, and there is plenty of math notations and equations, so I would suggest the following if you already have some pretty solid bases in statistics. However, the authors do use a language that can be understood by a large audience:

Pagano, M., & Gauvreau, K. (2018). *Principles of biostatistics* (2nd ed.). Chapman and Hall/CRC. ISBN 9781138593145.
Azen, R., & Walker, C. M. (2021). *Categorical data analysis for the behavioral and social sciences* (2nd ed.). Routledge. ISBN 9780367352769.

On the Simpson's paradox, see:

Hernán, M. A., Clayton, D., & Keiding, N. (2011). The Simpson's paradox unraveled. *International Journal of Epidemiology*, 40(3), 780–785. https://doi.org/10.1093/ije/dyr041.
Selvitella, A. (2017). The ubiquity of the Simpson's Paradox. *Journal of Statistical Distributions and Applications*, 4(1). https://doi.org/10.1186/s40488-017-0056-5.

On the issue of the minimum sample size required by the Cochran–Mantel–Haenszel test, see:

Munoz, A., & Rosner, B. (1984). Power and sample size for a collection of 2 × 2 tables. *Biometrics*, 40(4), 995–1004. https://doi.org/10.2307/2531150.
Woolson, R. F., Bean, J. A., & Rojas, P. B. (1986). Sample size for case-control studies using Cochran's statistic. *Biometrics*, 42(4), 927932. https://doi.org/10.2307/2530706.
Wittes, J., & Wallenstein, S. (1987). The power of the Mantel-Haenszel test. *Journal of the American Statistical Association*, 82(400), 1104–1109. https://doi.org/10.2307/2289387.
Nam, J. (1992). Sample size determination for case-control studies and the comparison of stratified and unstratified analyses. *Biometrics*, 48(2), 389–395. https://doi.org/10.2307/2532298.
Agresti, A. (1996). *An introduction to categorical data analysis*. John Wiley & Sons, Inc. ISBN 9780471113386.

On the combined use of stratum-specific chi-squared tests and the Cochran–Mantel–Haenszel test, in addition to the work by Azen and Walker cited earlier, see the following:

McDonald, J. H. (2008). *Handbook of biological statistics*. Sparky House Publishing.

On the limitations of the Breslow–Day test, it is important to note that for the test to be valid, the sample size should be relatively large in each stratum, and at least 80% of the expected cell counts should be greater than 5. For further discussions on the test, see the following pretty advanced works; the first and the last also compare the Breslow–Day to other competitor tests for the homogeneity of ORs across strata. The book published by Agresti in 1996 (previously cited) touches upon alternative methods for small-sample sizes:

Jones, M. P., O'Gorman, T. W., Lemke, J. H., & Woolson, R. F. (1989). A Monte Carlo investigation of homogeneity tests of the odds ratio under various sample size configurations. *Biometrics*, 45(1), 171–181. https://doi.org/10.2307/2532043.

Breslow, N. E. (1996). Statistics in epidemiology: The case-control study. *Journal of the American Statistical Association*, 91(433), 14–28. https://doi.org/10.1080/01621459.1996.10476660.

Lachin, J. M. (2010). *Biostatistical methods: The assessment of relative risks* (2nd ed.). John Wiley & Sons. ISBN 9780470508220.

Bagheri, Z., Ayatollahi, S. M. T., & Jafari, P. (2011). Comparison of three tests of homogeneity of odds ratios in multicenter trials with unequal sample sizes within and among centers. *BMC Medical Research Methodology*, 11(1). https://doi.org/10.1186/1471-2288-11-58.

Almalik, O., & Heuvel, E. R. van den. (2018). Testing homogeneity of effect sizes in pooling 2 × 2 contingency tables from multiple studies: A comparison of methods. In Y. Hu (Ed.), *Cogent Mathematics & Statistics*, 5(1), 478698. https://doi.org/10.1080/25742558.2018.1478698.

For the weighted average of Cramér's *V*, see:

Sakoda, J. M. (1977). Measures of association for multivariate contingency tables. In *Social statistics section proceedings for the American Statistical Association (Part III)*, 777-780.

For the use of the ratio of conditional odds ratios:

Heilbron, D. C. (1981). The analysis of ratios of odds ratios in stratified contingency tables. *Biometrics*, 37(1), 55–66. https://doi.org/10.2307/2530522.

Nay, W., Brown, R., & Roberson-Nay, R. (2013). Longitudinal course of panic disorder with and without agoraphobia using the national epidemiologic survey on alcohol and related conditions (NESARC). *Psychiatry Research*, 208(1), 54–61. https://doi.org/10.1016/j.psychres.2013.03.006.

Limoncin, E., Gravina, G. L., Corona, G., Maggi, M., Ciocca, G., Lenzi, A., & Jannini, E. A. (2017). Erectile function recovery in men treated with phosphodiesterase type 5 inhibitor administration after bilateral nerve-sparing radical prostatectomy: A systematic review of placebo-controlled randomized trials with trial sequential analysis. *Andrology*, 5, 863–872. https://doi.org/10.1111/andr.12403.

Niehoff, N., White, A. J., McCullough, L. E., Steck, S. E., Beyea, J., Mordukhovich, I., … Gammon, M. D. (2017). Polycyclic aromatic hydrocarbons and postmenopausal breast cancer: An evaluation of effect measure modification by body mass index and weight change. *Environmental Research, 152,* 17–25. https://doi.org/10.1016/j.envres.2016.09.022.

On the Bonferroni correction, see:

Simes, R. J. (1986). An improved Bonferroni procedure for multiple tests of significance. *Biometrika, 73*(3), 751–754. https://doi.org/10.1093/biomet/73.3.751

Nakagawa, S. (2004). A farewell to Bonferroni: The problems of low statistical power and publication bias. *Behavioral Ecology, 15*(6), 1044–1045. https://doi.org/10.1093/beheco/arh107.

Starnes, D. S., Yates, D., & Moore, D. S. (2011). *The practice of statistics* (4th ed.). W.H. Freeman. ISBN 9781429245593.

Armstrong, R. A. (2014). When to use the Bonferroni correction. *Ophthalmic and Physiological Optics, 34*(5), 502–508. https://doi.org/10.1111/opo.12131.

VanderWeele, T. J., & Mathur, M. B. (2018). Some desirable properties of the Bonferroni correction: Is the Bonferroni correction really so bad? *American Journal of Epidemiology, 188*(3), 617–618. https://doi.org/10.1093/aje/kwy250.

On another method (namely, Sidak's) to account for multiple comparisons in the context of cross-tab analysis, see:

Beasley, T. M., & Schumacker, R. E. (1995). Multiple regression approach to analyzing contingency tables: Post hoc and planned comparison procedures. *The Journal of Experimental Education, 64*(1), 79–93. https://doi.org/10.1080/00220973.1995.9943797.

7.4.7 Essential Readings for Chapter 7

On the advanced methods covered in this same chapter, the book by Azen and Walker previously cited features chapters on log-linear modelling and logistic regression. The latter method is described, also from the perspective of cross-tabs analysis, in the work by Allison cited below. The book by Hosmer and colleagues provides an in-dept general treatment of the topic. The works authored by Beh and Lombardo and by Greenacre are essential readings in case you want to know more about correspondence analysis. The book by Clausen features a section on the combined use of correspondence analysis and log-linear modelling:

Clausen, S. E. (1998). *Applied correspondence analysis.* SAGE Publications, Inc. ISBN 9780761911159.

Allison, P. D. (2012). *Logistic regression using SAS: Theory and application* (2nd ed.). SAS Publishing. ISBN 9781599946412.

Hosmer, D. W., Jr., Lemeshow, S., & Sturdivant, R. X. (2013). *Applied logistic regression.* Wiley. https://doi.org/10.1002/9781118548387.

Beh, E. J., & Lombardo, R. (2021). *An introduction to correspondence analysis.* John Wiley
    & Sons. ISBN 9781119041948.
Greenacre, M. (2021). *Correspondence analysis in practice* (3rd ed.). Chapman & Hall/
    CRC. ISBN 9780367782511.

7.5 Parting Thoughts on Cross-Tab Analysis

As we come to the end of this book on cross-tab analysis, I hope you have
gained a better understanding of this powerful analytical tool. Our initia-
tion into the world of cross-tabulations set the stage. We learned the art of
simplifying the complexities of categorical data, using the Titanic dataset
as our guide. This foundational understanding was crucial, teaching us to
see stories in rows and columns, and to extract valuable insights from our
cross-tabs.

Next, we delved into the relationships between nominal categorical variables by introducing the concept of independence. Through the lens of
the chi-squared test, we explored the relation between two variables. The
Titanic dataset once again served as our compass, revealing the dependence
between GENDER and SURVIVAL.

Our journey into the chi-squared test deepened as we tackled the chi-
squared residuals. Here, we learned the importance of understanding deviations and the stories they tell. The relationship between the chi-squared test
and sample size emerged as an important lesson, and equally significant (no
pun intended!) was our exploration of the chi-squared test and the issue of
small expected frequencies, a reminder of the care and precision required in
our analyses.

The chapter on the measures of association enriched our toolkit. We were
introduced to a body of measures, each offering a unique perspective on
the strength and direction of the dependence. From chi-squared-based measures, to PRE measures, to the ORs and Yule's Q, we learned to articulate the
depth and nuances of our data narratives.

Our journey into stratified 2 × 2 cross-tabs was akin to navigating a labyrinth, where each turn reveals a new perspective. This journey taught us
that data, much like stories, can have multiple layers, each offering a unique
perspective. The introduction of a third variable, such as the passenger
CLASS in the Titanic dataset, transformed our two-dimensional view into
a three-dimensional exploration. These stratified tables, reminiscent of layered cakes, allowed us to cross-tabulate two variables, like GENDER and
SURVIVAL, for each level or stratum of a third variable, such as CLASS.

In this stratified landscape, we encountered the concept of partial (or conditional) tables. These tables, like individual chapters of a book, told stories
specific to each stratum. For instance, our look at the relationship between

ABORTION OPINION and RELIGIOSITY, stratifying by GENDER, revealed nuanced patterns that might have been obscured in an aggregated view. Such stratification underscored the importance of context, showing us that overarching patterns might sometimes mask subtler, yet significant, relationships within specific strata. This dive into stratified cross-tabs emphasised that to truly understand a relationship, one must often dissect the data, peeling back its layers to uncover the complete story. By doing so, we learned that sometimes interesting insights lie hidden, waiting to be discovered through meticulous exploration.

Tying all these threads together, we were equipped with a comprehensive understanding, ready to apply our knowledge to other real-world scenarios. The book's culmination in a worked example offered a hands-on experience in reporting cross-tab analysis results in a succinct, yet informative, fashion.

As we expanded our horizons beyond 2 × 2 tables, we had a peek at the world of larger and more complex cross-tabs. While many of the methodologies and principles elucidated in this book provide a foundational understanding applicable to more complex datasets, embracing advanced methods becomes imperative when dealing with larger or multi-stratified tables. Methods like log-linear modelling, correspondence analysis, or logistic regression are bound to comes into play.

In this broader realm, while advanced techniques can proffer richer insights, they come with their own sets of challenges. The core of this book, anchored in the simplicity of 2 × 2 tables, provides a robust foundation upon which to build. As you delve deeper into more complex analyses, the fundamental lessons and methods covered here are bound to serve as your guide, assisting you in navigating complex data analyses with some confidence.

The field of categorical data and cross-tabs analysis is broad, but with the knowledge from this book you are hopefully better prepared to tackle it. I appreciate your time and effort in joining me through this content.

Best of luck in your future data endeavours!

Appendix: Worked Examples of Odds and Odds Ratios

Introduction

The purpose of this Appendix is to supplement the core content of the book with additional practical, hands-on examples that illuminate the concepts of odds and odds ratios, covered in Chapter 4. These statistical measures are pivotal in the analysis of cross-tabulated data, especially when the goal is to understand the patterns of association in our cross-tabs, in terms of strength and direction of the dependence between the levels of the variables under analysis.

Odds and odds ratios can be abstract and baffling concepts for those encountering them for the first time. While the main text of this book provides a theoretical foundation as well as practical examples of their calculation and interpretation, this Appendix aims to reiterate these concepts to solidify the reader's mastery, which is also crucial for understanding advanced techniques like logistic regression, introduced in Chapter 7. By working through the following examples, readers will gain a deeper understanding of how odds and odds ratios are calculated, interpreted, and applied in new scenarios.

Each example is structured to first present the scenario and the data, followed by a step-by-step walkthrough of the calculations. We then delve into the interpretation of the results, discussing what they mean in the context of the research question at hand.

The examples chosen reflect a variety of scenarios, all situated within a criminal justice framework that examines factors influencing the rehabilitation outcomes of ex-offenders. These scenarios are designed to address potential research questions such as the effectiveness of intervention programmes, the impact of educational attainment on recidivism, and the role of gender in behavioural outcomes. They are designed to progressively build the reader's skills and confidence in working with these measures.

By the end of this Appendix, readers should feel comfortable tackling their own analyses of odds and odds ratios, armed with the knowledge and practice they have gained throughout the entire book.

A.1 Reoffending Status and Program Completion

Scenario and Data:

This example explores the relationship between REOFFENDING STATUS and PROGRAM COMPLETION among a group of ex-offenders (see also example A.5). The data is cross-tabulated to show the frequency of reoffending for those who completed the program versus those who did not (Table A.1).

Individual Odds and Interpretation:

(a) Odds Reoffending Status "Yes" | Program Completion "No" 109 : 98 = 1.11 : 1 = 1.11

For those who did not complete the program, the odds of reoffending are a bit above 1, indicating a slightly higher likelihood of reoffending than not reoffending.

(b) Odds Reoffending Status "Yes" | Program Completion "Yes" 34 : 159 = 1 : 4.68 = 0.213

For those who completed the program, the odds of reoffending are 0.213, which indicates a substantially lower likelihood of reoffending compared to not reoffending.

Odds Ratio and Interpretation:

$$OR\ 1.11 : 0.213 = 5.20$$

An OR greater than 1 suggests that the odds are higher for the first group. In this context, the OR of 5.20 means that individuals who did not complete the program are about 5 times more likely to reoffend compared to

TABLE A.1

Cross-Tabulation of REOFFENDING STATUS and PROGRAM COMPLETION for a Hypothetical Group of 400 Individuals Undergoing Rehabilitation

	PROGRAM COMPLETION		
REOFFENDING STATUS	**No**	**Yes**	**Total**
No	98	159	257
Yes	109	34	143
Total	207	193	400

Note: Chi-squared: 53.4; df: 1; p-value: <0.01; V: 0.365.

those who did complete the program. This is a sizeable difference and suggests that program completion has a strong protective effect against reoffending.

Overall Takeaway:
Program completion appears to be effective in reducing the likelihood of reoffending. The odds of reoffending are over 5 times higher for those who did not complete the program compared to those who did, making this an area of focus for reducing reoffending. There is a statistically significant association between REOFFENDING STATUS and PROGRAM COMPLETION, as the chi-squared test indicates (chi-squared: 53.4; df: 1; p-value: <0.01; V: 0.365).

A.2 Program Completion and Educational Level

Scenario and Data:
In this example, we examine the relationship between the PROGRAM COMPLETION and the EDUCATIONAL LEVEL of participants. The data is presented in a table showing the number of individuals who completed and did not complete the program across three education levels: "None", "High School", and "College" (Table A.2).

Individual Odds and Interpretation:

  (a) Odds Program Completion "No" | Educational Level "None" 36 : 32
      = 1.125 : 1 = 1.13
  (b) Odds Program Completion "No" | Educational Level "High School"
      137 : 124 = 1.10 : 1 = 1.10
  (c) Odds Program Completion "No" | Educational Level "College" 34 :
      37 = 1 : 1.09 : 1: = 0.92

TABLE A.2

Cross-Tabulation of PROGRAM COMPLETION and EDUCATIONAL LEVEL for a Hypothetical Group of 400 Individuals Undergoing Rehabilitation

	PROGRAM COMPLETION		
EDUCATIONAL LEVEL	**No**	**Yes**	**Total**
None	36	32	68
High School	137	124	261
College	34	37	71
Total	207	193	400

Note: Chi-squared: 0.520; df: 2; p-value: 0.771.

The odds of not completing the program are slightly greater than 1 for the "None" and "High School" levels, suggesting a slightly higher likelihood of not completing the program compared to completing it. The odds of not completing the program are smaller than 1 for the "College" level, which indicates that who has college education is slightly less likely to not complete the program (1 : 0.92 = 1.09).

Odds Ratios and Interpretation:

$$\text{OR a–b } 1.13 : 1.10 = 1.03$$

Individuals with no formal education are marginally more likely to not complete the program compared to those with a high school education. The OR marginally larger than 1 indicates that there is no sizeable difference in program completion rates between these two groups.

$$\text{OR a–c } 1.13 : 0.92 = 1.23$$

Individuals with no formal education are 1.23 times more likely to not complete the program compared to those with a college education.

Overall Takeaway:
The odds of not completing the program are somewhat higher for the "None" and "High School" education levels compared to the "College" level. However, the ORs across different educational levels are close to 1, indicating only minor differences in program completion rates based on educational background. These differences are not substantial, and the chi-squared test (chi-squared: 0.520; df: 2; p-value: 0.771) confirms that there is no statistically significant association between EDUCATIONAL LEVEL and PROGRAM COMPLETION.

A.3 Employment Status and Educational Level

Scenario and Data:
This example investigates the association between EMPLOYMENT STATUS post-release and EDUCATIONAL LEVEL of ex-offenders (Table A.3). The cross-tabulation includes the number of employed and unemployed individuals across the different education levels taken into account in the previous example.

Individual Odds and Interpretation:

(a) Odds Employment Status "Unemployed" | Educational Level "None" 42 : 26 = 1.62 : 1 = 1.62

TABLE A.3

Cross-Tabulation of EMPLOYMENT STATUS and EDUCATIONAL LEVEL for a Hypothetical Group of 400 Individuals Undergoing Rehabilitation

	EDUCATIONAL LEVEL			
EMPLOYMENT STATUS	None	High School	College	Total
Employed	26	128	52	206
Unemployed	42	133	19	194
Total	68	261	71	400

Note: Chi-squared: 18.9; df: 2; *p*-value: <0.001; *V*: 0.217.

Individuals without any education are about 1.6 times more likely to be unemployed than to be employed. This indicates a higher likelihood of unemployment among this group.

(b) Odds Employment Status "Unemployed" | Educational Level "High School" 133 : 128 = 1.04 : 1 = 1.04

The likelihood of being unemployed versus employed is nearly balanced for those with a high school education. The odds are slightly tilted towards unemployment but not by much.

(c) Odds Employment Status "Unemployed" | Educational Level "College" 19 : 52 = 1 : 2.74 = 0.37

College-educated individuals are substantially less likely to be unemployed compared to being employed. In fact, they are about 2.7 times more likely to be employed than unemployed.

Odds Ratios and Interpretation:

OR a–b 1.62 : 1.04 = 1.56

Individuals with no education are approximately 1.56 times more likely to be unemployed compared to those with a high school education. This tells us that the odds of unemployment are higher for those with no formal education compared to those who have completed high school.

OR a–c 1.62 : 0.37 = 4.38

Individuals with no education are roughly 4.4 times more likely to be unemployed than those with a college education. This highlights a sizeable disparity in unemployment odds between the least and most educated groups in the dataset.

Overall Takeaway:

The ORs suggest that educational attainment plays an important role in employment status. Specifically, having no education substantially increases one's odds of being unemployed, especially when compared to having a college education. The OR between "None" and "High School" is also sizeable, but less so than between "None" and "College" degree. Overall, the association between EMPLOYMENT STATUS and EDUCATIONAL LEVEL is statistically significant, as the chi-squared test indicates (chi-squared: 18.9; df: 2; p-value: <0.001; V: 0.217).

A.4 Behaviour Rating and Gender

Scenario and Data:

This example investigates the association between GENDER and BEHAVIOUR RATING of ex-offenders. The behaviour rating is categorised as "Low", "Medium", and "High", and the participants are divided by gender (Table A.4).

Individual Odds and Interpretation:

(a) Odds Behaviour Rating "Low" | Gender "Male" 84 : 26 = 3.23 : 1 = 3.23

Males are approximately 3.2 times more likely to have a low behavioural rating compared to females. This suggests that low behavioural ratings are more common among males than females.

(b) Odds Behaviour Rating "Medium" | Gender "Male" 101 : 31 = 3.26 : 1 = 3.26

TABLE A.4

Cross-Tabulation of BEHAVIOUR RATING and GENDER for a Hypothetical Group of 400 Individuals Undergoing Rehabilitation

BEHAVIOUR RATING	GENDER		
	Female	Male	Total
Low	26	84	110
Medium	31	101	132
High	103	55	158
Total	160	240	400

Note: Chi-squared: 69.0; df: 2; p-value: <0.001; V: 0.415.

Males are around 3.3 times more likely to have a medium behavioural rating compared to females. Like the low rating, medium behavioural ratings also seem to be more common among males than females.

(c) Odds Behaviour Rating "High" | Gender "Male" 55 : 103 = 1 : 1.82 = 0.53

Males are less likely to have a high behavioural rating compared to females. In fact, females are about 1.8 times (1 : 0.53) more likely to have a high behavioural rating than males.

Odds Ratios and Interpretation:

$$OR\ a–b\ 3.23 : 3.26 = 0.99$$

The odds of males having a low behavioural rating are almost the same as the odds of them having a medium rating when compared to females (OR close to 1). This suggests that among males, the likelihood of having a low or medium behavioural rating is approximately equal when gender is taken into account.

$$OR\ a–c\ 3.23 : 0.53 = 6.09$$

Males are about 6.1 times more likely to have a low behavioural rating compared to a high behavioural rating when compared to females. This suggests that the likelihood of males having a low rating is considerably greater than those having a high rating in comparison to females.

Overall Takeaway:
The first OR indicates that, when compared to females, the odds of males having a low behavioural rating are nearly identical to the odds of them having a medium rating. The second OR tells us that males are substantially more likely to have a low behavioural rating than a high one, compared to females. These ORs confirm and quantify the disparities suggested by the individual odds. Males are notably more likely to have lower behavioural ratings (either "Low" or "Medium") compared to "High" ratings when gender is factored in. The OR between "Low" and "Medium" being close to 1 also validates that these two categories are similarly common among males when compared to females. The association between BEVAHIOUVAL RATING and GENDER is statistically significant (chi-squared: 69.0; df: 2; p-value <0.001; V: 0.415).

A.5 Gender and Reoffending Status, Controlling for Program Completion

Scenario and Data:
In this analysis, we investigate the association between GENDER and REOFFENDING STATUS, while controlling for PROGRAM COMPLETION. The data is organised into a three-way table (2 × 2 × 2),

which cross-tabulates REOFFENDING STATUS and GENDER, stratified by PROGRAM COMPLETION. This example is designed to investigate the extent to which the stratifying variable may affect the relationship between the other two (Table A.5).

Individual Conditional Odds and Interpretation:
Program Completion "No"

(a) Odds Reoffending Status "No" | Gender "Female" 50 : 34 = 1.47 : 1 = 1.47

Females who did not complete the program are about 1.5 times more likely to not reoffend than to reoffend. This suggests a higher likelihood of avoiding reoffence among the females.

(b) Odds Reoffending Status "No" | Gender "Male" 48 : 75 = 1 : 1.56 = 0.64

Males who did not complete the program are less likely to avoid reoffending compared to females. In fact, they are 1.56 times more likely to reoffend than not.
Program Completion "Yes"

(c) Odds Reoffending Status "No" | Gender "Female" 70 : 6 = 11.6 : 1 = 11.6

TABLE A.5

Cross-Tabulation of REOFFENDING STATUS and GENDER, Stratified by PROGRAM COMPLETION, for a Hypothetical Sample of 400 Individuals Undergoing Rehabilitation

		REOFFENDING STATUS		
PROGRAM COMPLETION	GENDER	No	Yes	Total
No	Female	50	34	84
	Male	48	75	123
	Total	98	109	207
Yes	Female	70	6	76
	Male	89	28	117
	Total	159	34	193
Total	Female	120	40	160
	Male	137	103	240
	Total	257	143	400

Note: The first two sections represent partial (conditional) tables for those who did not complete and those who completed the program, detailing counts of reoffence by GENDER. The final section presents the marginal totals, providing an aggregate view across both levels of PROGRAM COMPLETION (CMH statistic: 16.15; df: 1; p-value: <0.001; BD statistic: 0.713; df: 1; p-value: 0.398).

Females who completed the program are substantially more likely to not reoffend, being about 11.6 times more likely to avoid reoffending than to reoffend.

(d) Odds Reoffending Status "No" | Gender "Male" 89 : 28 = 3.18 : 1 = 3.18

Males who completed the program are also more likely to not reoffend, but to a lesser extent compared to females, being about 3.2 times more likely to avoid reoffending than to reoffend.

Conditional Odds Ratios and Interpretation:

$$OR\ a\text{–}b\ 1.47 : 0.64 = 2.30$$

Females who did not complete the program are 2.3 times more likely to not reoffend compared to males who also did not complete the program. This indicates a gender difference in reoffence among those who did not complete the program.

$$OR\ c\text{–}d\ 11.6 : 3.18 = 3.65$$

Females who completed the program are about 3.65 times more likely to not reoffend compared to males who completed the program. This also indicates a gender difference in reoffence among those who did complete the program.

Individual Marginal Odds and Interpretation:

(e) Odds Reoffending Status "No" | Gender "Female" 120 : 40 = 3 : 1 = 3

The odds of 3 for females overall suggest that they are three times more likely to not reoffend than to reoffend.

(f) Odds Reoffending Status "No" | Gender "Male" 137 : 103 = 1.33 : 1 = 1.33

The odds of 1.33 for males overall indicate that they are somewhat more likely to not reoffend than to reoffend, but the likelihood is not as strong as it is for females.

Marginal Odds Ratio and Interpretation:

$$OR\ e\text{–}f\ 3 : 1.33 = 2.26$$

Overall, females are about 2.26 times more likely to not reoffend compared to males. This marginal odds ratio highlights a significant gender disparity in reoffence across the entire sample.

Overall Takeaway:
The marginal OR reveals a general trend of females being more likely to not reoffend compared to males. The association is significant, as indicated by the chi-squared test (chi-squared: 13.42; df: 1; p-value: <0.001; V: 0.183). However, this observation does not consider the role of PROGRAM COMPLETION.

Stratifying by PROGRAM COMPLETION, the GENDER-REOFFENDING STATUS association persists across both strata, suggesting a consistent pattern of association (first partial table, chi-squared: 8.41, df: 1, p-value: <0.01, V: 0.202; second partial table, chi-squared: 8.17, df: 1, p-value: <0.01, V: 0.206; CMH statistic: 16.15, df: 1, p-value: <0.001).

Further examination of conditional ORs reveals that females, whether they completed the program or not, have higher odds of not reoffending compared to their male counterparts. Notably, the BD test (test statistic: 0.713; df: 1; p-value: 0.398) shows homogeneity in ORs across strata, implying that PROGRAM COMPLETION does not significantly alter the GENDER-REOFFENDING STATUS association. The MH common odds ratio estimate of 2.65 indicates that females are, on average, more than twice as likely not to reoffend as males, regardless of program completion.

In summary, the analysis strongly suggests that GENDER plays a significant role in reoffending likelihood, a relationship that is not substantially influenced by whether participants complete the rehabilitation program.

Final Reflections

As we conclude this Appendix, we reflect on the journey through the interesting world of cross-tab analysis. These examples, ranging from exploring the dynamics of programme completion to understanding the influence of educational levels on rehabilitation outcomes, have served to demystify the concepts of odds and odds ratios, making them more tangible. Through this exploration, we have seen how these measures are not mere abstractions but powerful tools for uncovering insights in real-world data. Together with formal tests, they enable us to quantify the strength and direction of associations in a meaningful way, providing a clearer understanding of the factors that shape patterns of association in our cross-tabs.

This Appendix, complementing the core content of the book, aimed to solidify your grasp of these concepts, crucial for any data-driven analysis. Whether you are a student, a researcher, or a practitioner, the skills and understanding you have gained here are foundational tools. They empower you to conduct your analyses with some confidence and interpret statistical results with a more nuanced perspective. Remember, the true power of statistics lies in its ability to illuminate patterns and relationships that might otherwise remain hidden. As you close this Appendix (and this book), carry forward the knowledge and insights you have gained, apply them to your own research and practice, and continue to explore the fascinating world of categorical data analysis with curiosity and, more importantly, rigour.

Glossary of Terms

Average Expected Count Rule: Guideline used in the analysis of cross-tabulations for the chi-squared test. It states that for the chi-squared test to be reliable at the 0.05 significance level, the average expected frequency across all cells in a cross-tab should be at least 5 or 6. This average is calculated by dividing the table's grand total by the number of cells (number of rows multiplied by the number of columns). For example, in a 2 × 2 table, the average expected frequency is obtained by dividing the grand total by 4 (2 rows times 2 columns). The rule helps to determine whether a given dataset is suitable for a chi-squared test, ensuring that the test's results are statistically valid. It also enables the estimation of a minimum grand total necessary for a cross-tab to maintain chi-squared test's reliability, calculated as 5 or 6 times the total number of cells (Chapter 3). See also: Chi-Squared Test, Expected Frequencies, Minimum Grand Total for Chi-Squared Test Reliability.

Bonferroni Correction: A method to adjust the level of significance when one conducts multiple tests or comparisons. The original significance level is divided by the number of tests performed. This results in a new stricter threshold that each individual test must meet to be considered statistically significant. This adjustment aims to reduce the risk of incorrectly identifying an association or effect as significant merely due to random chance (Chapter 5). See also: *p*-value.

Breslow–Day's Test: A statistical test used to assess the homogeneity of the odds ratios across different levels of a third variable in stratified 2 × 2 tables. It tests the null hypothesis that the odds ratios are equal across all strata (Chapter 5). See also: Homogeneous Association.

Categorical Variable: Variables that deal with non-numeric data, sorting observations into specific categories. They are further divided into nominal (without order) and ordinal (with a distinct order) (Chapter 1). See also: Nominal Variable, Ordinal Variable.

Chi-Squared Test: A statistical test used to determine if there is a significant association between two categorical variables. The test statistic is calculated by summing the squared differences between observed and expected counts in each cell of a cross-tabulation, divided by the expected counts. This sum follows a chi-square distribution under the null hypothesis of no association. The resulting test statistic is compared against that distribution to determine the *p*-value, which indicates the probability of observing such a statistic under the null hypothesis (Chapter 2). See also: Expected Frequencies, Minimum

Grand Total for Chi-Squared Test Reliability, Observed Frequencies, *p*-Value.

Cochran–Mantel–Haenszel's Test: A statistical test used in the analysis of $2 \times 2 \times K$ tables, where K is the number of levels of the stratifying variable. It is used to test for an association between two categorical variables while controlling for a third categorical one (Chapter 5). See also: Conditional Independence.

Cohen's Thresholds: Guidelines established by statistician J. Cohen to interpret the strength of association expressed by chi-squared-based coefficients. These benchmarks categorise the strength of the association as small, medium, or large, and are scaled according to the degrees of freedom (df) associated with the table. In this context, the df is the lesser of the number of rows or columns in the contingency table minus one. Cohen's thresholds are instrumental for comparing the magnitude of associations across studies with varying table sizes. They serve as a reference to gauge the practical significance of findings, complementing the statistical significance and aiding in the interpretation of chi-squared-based measures of association. They should be used with caution, considering the specific circumstances and domain of the research (Chapter 4). See also: Chi-Squared Test, Measures of Categorical Association.

Conditional Independence: In the context of stratified cross-tabulation analysis, conditional independence describes a situation where two variables are independent of each other, specifically when the effect of a third variable is controlled or held constant. This means that any association between the first two variables disappears or becomes statistically insignificant when this third variable is accounted for. The concept is crucial for isolating the effects of different variables, helping to determine if a relationship between two variables is influenced by another variable (Chapter 5). See also: Homogeneous Association, Independence, Stratified 2×2 Cross-Tabs.

Contingency Coefficient (C): A measure to quantify the strength of the association between two categorical variables. It is calculated by dividing the chi-squared value by itself plus the table's grand total, and then taking the square root. Its value ranges between 0.0 (indicating independence) and less than 1.0 (Chapter 4).

Contingency Coefficient Adjusted (C_{adj}): Adjusted version of C, ensuring its value is bounded between 0.0 and 1.0, inclusive. It facilitates comparisons between tables of different sizes, addressing the limitation of C not reaching its maximum value of 1.0 (Chapter 4).

Contingency Table: see Cross-Tabulation.

Correspondence Analysis: A statistical technique for visualising relationships within cross-tabs of different types. CA provides a means of displaying or summarising a set of data in two-dimensional graphical form by transforming the association present in contingency tables into geometric relationships, thus enabling easier interpretation of complex associations within the data (Chapter 7).

Cramér's *V*: A measure of association for categorical variables; it is calculated from the chi-squared statistic, normalised by the sample size and the lesser of the table's dimensions minus one. This measure is suitable for tables larger than 2 × 2 and is bounded between 0.0 (no association) and 1.0 (perfect association). However, it has limitations, particularly with tables that have a concentration of observations in specific cells, sparse cells with very low or zero counts, or uneven marginal frequencies. In these conditions, *V* can overestimate the strength of the association. For example, a table with observed counts close to expected ones across most cells, except for a single cell with a small observed and expected count, can result in a high *V* value, suggesting a strong association where there might be little to none. An alternative measure, the *W* coefficient, may provide a more accurate reflection of association in such cases, adjusting for the limitations of *V* (Chapter 4). See also: Measures of Categorical Association.

Cramér's *V* (Weighted Average of): Weighted average of *V* across strata; it measures the strength of the association between two variables while controlling for a third one (Chapter 5). See also: Stratified 2 × 2 Cross-Tabs.

Cross-Tabulation (Cross-Tab): A table showing the count of observations at each intersection of two categorical variables' levels, revealing the frequency and patterns of combinations in the data (Chapter 1).

Expected Frequencies (Expected Counts): Theoretical counts indicating how many individuals would be expected to fall at the intersections of the levels of the two variables in a cross-tabulation under the null hypothesis of no association. These counts are calculated based on the product of the marginal totals of each variable, divided by the table's grand total. These counts provide a baseline for comparing observed frequencies, which is essential in determining if the observed data significantly deviate from what would be expected under the null hypothesis (Chapter 2). See also: Chi-Squared Test, Observed Frequencies.

Fisher's Test: A statistical method used for 2 × 2 tables, especially with small expected frequencies. Unlike the chi-squared test, it calculates the exact probability of observing a specific table under the assumption of no association, using the hypergeometric distribution. It provides a conservative assessment ideal for cross-tabs featuring small expected counts (smaller than 1). Named after British statistician Ronald A. Fisher (Chapter 3). See also: $(N-1)/N$ Corrected Chi-Squared Statistic.

Goodman–Kruskal's λ (Lambda): A measure of association based on the idea of Proportional Reduction in Error (PRE). It is bounded between

0.0 and 1.0. For example, a λ of 0.42 can be interpreted by saying that the knowledge of variable A improves our ability to predict variable B by 42% (Chapter 4). See also: Measures of Categorical Association.

Haldane–Anscombe Correction: Adjustment applied to 2 × 2 tables when calculating the odds ratio in cross-tabs containing zeros in any of its diagonals. The correction entails adding 0.5 to every cell in the table before determining the odds ratio (Chapter 4).

Homogeneous Association: A situation where the strength of the association between two variables is consistent across the levels of a third variable (Chapter 5). See also: Breslow–Day's Test, Conditional Independence, Stratified 2 × 2 Cross-Tabs.

Independence: In the context of cross-tabulation analysis, independence refers to the assumption that the variables being analysed do not have any significant association with each other. When two variables are independent, the distribution of one variable does not affect the distribution of the other variable. In a cross-tabulation, independence is supported when the observed counts in the cells are close to the expected counts under the assumption of no association between the variables (Chapter 2). See also: Conditional Independence.

Interaction in Stratified Analysis: This concept describes a situation where the relationship between two primary variables in a cross-tab changes or varies when analysed across different levels of a third variable. In stratified analysis, interaction (or three-way association) suggests that the association between the primary variables is not consistent (in terms of strength and/or direction) across the various strata or groups defined by the third variable. This variation highlights the dynamic and complex nature of relationships in datasets, emphasising the need to consider how different layers can influence the observed associations (Chapter 5). See also: Breslow–Day's Test, Conditional Independence, Homogeneous Association, Stratified 2 × 2 Cross-Tabs.

Interpretation of Association Measures: The process of evaluating the strength and direction of the relationship between two variables using various statistical measures. This involves contextualising the numerical value of an association coefficient within the framework of established guidelines or thresholds, such as Cohen's thresholds, to articulate the relationship's strength in a meaningful way. It is an important step in the analysis of contingency tables, ensuring that the results are not only statistically significant but also practically significant (Chapter 4). See also: Cohen's Thresholds, Measures of Categorical Association.

Log-Linear Modelling: A statistical method that allows for the analysis of interactions between variables within complex, multidimensional contingency tables, providing insights into patterns and associations across many categorical variables (Chapter 7).

Logistic Regression: A statistical method used for predicting the outcome of a categorical dependent variable, based on one or more independent variables, which can also be categorical. The dependent variable can feature two (Yes/No, True/False, Dead/Alive) or more levels. Logistic regression describes the relationship between the dependent variable and the predictors, enabling the estimation of the probability of a particular outcome based on the given predictors. It also illuminates the independent effects of each variable, as well as possible interactions (Chapter 7).

Mantel–Haenszel Estimate: A method used to estimate a common odds ratio across different levels of a third variable in stratified 2 × 2 tables when the hypothesis of homogeneous association holds. It provides a single summary measure of the common odds ratio, adjusting for the effect of a stratifying variable (Chapter 5). See also: Homogeneous Association, Stratified 2 × 2 Cross-Tabs.

Marginal Frequencies: The column sums and the row sums in a cross-tabulation (Chapter 1).

Measures of Categorical Association: Measures that quantify the strength and direction of the relationship between two categorical variables. These measures can be divided into two groups: chi-square-based (contingency coefficient C, Phi coefficient, Phi corrected, Cramér's V, W) and non-chi-square-based (Goodman–Kruskal's λ , Odds Ratio, Yule's Q) (Chapter 4).

Minimum Grand Total for Chi-Squared Test Reliability: Minimum total count required in a cross-tabulation for the chi-squared test to yield reliable results. The guideline suggests an average expected frequency of at least 5 or 6 across all cells in the table to maintain the test's reliability at the 0.05 significance level. The minimum grand total necessary for a cross-tab to achieve this average expected count can be determined by multiplying 5 or 6 by the total number of cells (number of rows multiplied by number of columns). For instance, a 2 × 2 table requires a minimum grand total of 20 or 24, and a 2 × 4 table needs at least 40 or 48 for reliable chi-squared test results (Chapter 3). See also: Average Expected Count Rule, Chi-Squared Test, Expected Frequencies.

Monte Carlo Test of Independence: A statistical method for assessing the association between two categorical variables when the traditional chi-squared test's assumptions are violated, such as when expected cell counts are too small. Unlike the permutation approach, which reshuffles the observed data within the cells of the table, the Monte Carlo method generates completely new datasets based on the observed marginal totals. This method generates an empirical distribution of the chi-squared statistic under the null hypothesis of independence. Multiple random samples are generated, and the chi-squared statistic is computed for each. The p-value is then estimated as the proportion of simulated chi-squared values that are at least as extreme as the observed one (Chapters 2 and 3). See also: Chi-Squared Test, Fisher's Test, Permutation Test of Independence, *p*-value.

Mosaic Plot: A visual tool representing relationships between two categorical variables using rectangular tiles. Each tile's height shows the proportion of one category, its width indicates overall category proportion, and its colour reflects deviations from expected values in terms of (adjusted) standardised residuals. Useful for discerning patterns and deviations in larger tables (Chapter 3). See also: Standardised Chi-Squared Residuals.

Nominal Variable: Categorical variable that classifies without any inherent order (Chapter 1).

(N–1)/N Corrected Chi-Squared Statistic: A correction applied to the chi-squared statistic to reduce bias. It is useful for the analysis of cross-tabs where the grand total is smaller than 20 and the expected counts are equal to or larger than 1. It is calculated by multiplying the original chi-squared value by the factor $(N-1)/N$, where N is the sample size (table's grand total) (Chapter 3). See also: Fisher's Test.

Observed Frequencies (Observed Counts): The numbers indicating how many individuals fall at the intersections of the groups of the two variables in a cross-tabulation (Chapter 2). See also: Chi-Squared Test, Expected Frequencies.

Odds Ratio (OR): A measure used in the analysis of cross-tabs, specifically for quantifying the strength of dependence between pairs of categorical variables. An OR of 1 indicates independence between two variables. If OR > 1, there is a positive association, while OR < 1 indicates a negative association. It is not reliant on the chi-squared statistic (Chapter 4). See also: Measures of Categorical Association.

Ordinal Variable: Categorical variable that possesses an implicit order (Chapter 1).

Permutation Test of Independence: A statistical method used to determine the significance of the association between two categorical variables, particularly in larger contingency tables. It involves randomly shuffling, or permuting, the labels of one of the variables, while keeping the marginal totals fixed, to generate a distribution of the chi-squared statistic under the null hypothesis of independence. The *p*-value is then calculated as the proportion of permuted tables that yield a chi-squared statistic as extreme or more extreme than that of the observed table. This method provides an empirical alternative to traditional tests like Fisher's test and the chi-squared test, especially when expected frequencies are small and computational efficiency is desired. Unlike the Monte Carlo Test of Independence, which generates new datasets, the permutation test uses the original data to directly assess the significance within the existing dataset structure

(Chapter 3). See also: Chi-Squared Test, Fisher's Test, Monte Carlo Test of Independence, p-Value.

Phi Coefficient (Φ): A measure of association for 2×2 tables. It is the square root of the chi-squared statistic divided by the sample size. The Φ coefficient ranges from 0.0 (no association) to 1.0 (perfect association) (Chapter 4). See also: Measures of Categorical Association.

Phi Coefficient Corrected (Φ_{corr}): A normalised version of the ϕ coefficient. It accounts for the fact that the uncorrected ϕ does not always have 1.0 as maximum achievable value, as it depends on the configuration of the marginal sums. This, in turn, makes the uncorrected ϕ not directly comparable across tables with different marginals (Chapter 4).

PRE (Proportional Reduction in Error) Measures of Association: Measures of association used to quantify the strength of the relationship between two categorical variables based on the concept of proportional reduction in error. They help to indicate how much better we can predict one variable based on knowledge of another variable, compared to not having that knowledge. An example is Goodman–Kruskal's λ (Chapter 4).

p-Value: In the context of testing independence between two categorical variables, the p-value quantifies the evidence against the null hypothesis of independence. It represents the probability of observing a test statistic (chi-squared value) as extreme as the observed one, assuming the two variables are independent in the parent population. A common significance threshold is 0.05; a p-value ≤ 0.05 typically suggests sufficient evidence to reject the null hypothesis of independence, indicating a significant association (Chapter 2). See also: Bonferroni Correction, Chi-Squared Test, Monte Carlo Test of Independence, Permutation Test of Independence.

Quantitative Variables: Variables that represent measurable quantities and can be discrete (distinct values) or continuous (range values) (Chapter 1).

Raw Chi-Squared Residuals: Vales calculated by taking the difference between the observed count and the expected count in every cell of a cross-tabulation. Raw chi-squared residuals represent the unadjusted discrepancy between what is observed and what would be expected under the null hypothesis of independence (Chapter 3). See also: Standardised Chi-Squared Residuals.

Simpson's Paradox: A statistical phenomenon where an observed relationship in aggregate data disappears or reverses when data is stratified into subgroups. It underscores the importance of considering potential confounding variables and the value of stratifying data in cross-tab analyses to uncover underlying associations (Chapter 5).

Standardised Chi-Squared Residuals: They re-express the chi-squared raw residuals in terms of a z-score, helping to identify significant divergences from the hypothesis of independence. A value larger than 1.96 (in absolute terms) indicates a statistically significant dependence in that cell (Chapter 3).

Standardised Chi-Squared Residuals (Adjusted): A refinement of standardised residuals; they adjust for the row and column proportions of the cross-tab. Cells with absolute values larger than 1.96 indicate significant differences from what would be expected under independence (Chapter 3).

Statistical Variables: Characteristics or quantities that are observed or measured for each subject in a sample (Chapter 1).

Stratified 2 × 2 Cross-Tabs: Cross-tabulations that are broken down by levels of a third variable, allowing for the examination of how the relationship between two variables may change when conditioned on a third variable (Chapter 5). See also: Breslow–Day's Test, Cochran–Mantel–Haenszel's Test, Conditional Independence, Cramér's V (weighted average of), Homogeneous Association, Mantel–Haenszel Estimate.

Table's Grand Total: The overall number of individuals in a sample as represented in a cross-tabulation (Chapter 1).

Three-Way Association: See: Interaction in Stratified Analysis.

W Coefficient: A statistical coefficient, bounded between 0.0 and 1.0, developed to provide a more accurate representation of the strength of association in contingency tables, especially in cases where traditional measures like Cramér's V may be misleading due to concentration of observations, sparse cells, or uneven marginal frequencies. The calculation of the coefficient is more intricate than that of V and is designed to mitigate the impact of the table's size and distribution on the association measure. While the specifics of the W coefficient are beyond the scope of the book, it is noted for its utility in providing a more balanced assessment of association when traditional measures may be biased (Chapter 4). See also: Cramér's V, Measures of Categorical Association.

Yule's Q: A measure of association related to the OR only used for 2 × 2 contingency cross-tabs. It ranges from −1.0 (perfect negative association) to 1.0 (perfect positive association), with 0 indicating no association. Care is needed in interpretation, especially when the table contains zero frequencies (Chapter 4). See also: Odds Ratio.

Index

Page numbers in *italic* indicate figure and **bold** indicate table respectively

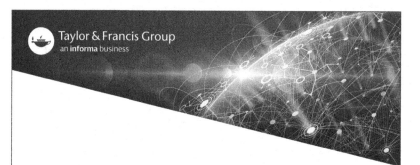

Printed in the United States
by Baker & Taylor Publisher Services

Printed in the United States
by Baker & Taylor Publisher Services